Troubleshooting and Repairing
VCRs

To my father, E.W. McComb, who always put back together what I took apart.

Troubleshooting and Repairing
VCRs

Gordon McComb

TAB BOOKS Inc.
Blue Ridge Summit, PA

FIRST EDITION
SEVENTH PRINTING

Copyright © 1988 by Gordon McComb
Printed in the United States of America

Library of Congress Cataloging in Publication Data

McComb, Gordon.
Troubleshooting and repairing VCRs / by Gordon McComb.
p.   cm.
Bibliography: p.
Includes index.
ISBN 0-8306-0060-4   ISBN 0-8306-2960-2 (pbk.)
1. Video tape recorders and recording—Maintenance and repair.
I. Title.
TK6655.V5M29   1988
621.388'332'0288—dc 19   87-33522
CIP

TAB BOOKS Inc. offers software for
sale. For information and a catalog,
please contact TAB Software Department,
Blue Ridge Summit, PA 17294-0850.

Questions regarding the content of this book
should be addressed to:

Reader Inquiry Branch
Editorial Department
TAB BOOKS Inc.
Blue Ridge Summit, PA 17294-0214

Contents

Acknowledgments   viii

Introduction   ix

Chapter 1   1
Introduction to Video Cassette Recorders
VCR Formats—VCR Types—Video and Audio Heads—Tuning—Operating Controls—Basic
Operation—The Enhanced VCR

Chapter 2   19
How VCRs Work
Television History 101—What Is Television?—Differences between Baseband and Rf
Signals — Basic VCR Block Diagram — Cassette Loading — Tape Threading — Tape
Transport—Recording and Playback Operations—Video Head Geometries—Inside Hi-fi
Audio—Super Beta—Super VHS—Digital Effects—PCM Audio—Controls—Connectors

Chapter 3   64
The VCR Environment
Unpacking—Installation—Hookup—Parts for Proper Hookup—Checkout—If Something Goes
Wrong—Achieving Optimum Picture and Sound—Proper Tape Handling—Video Accessories

Chapter 4   86
Tools and Supplies for VCR Maintenance
Workspace Area — Basic Tools — Volt-Ohmmeter — Logic Probe — Logic Pulser —
Oscilloscope—Frequency Meter—Infrared Detector—TV Set or Monitor—Assorted
Supplies—Capstan Roller Rejuvenator Lathe—Using Test Tapes

Chapter 5   107

General Cleaning and Preventative Maintenance

Frequency of Checkup—A Word of Caution—Personal Safety—General Maintenance—Internal Preventative Maintenance—Video Heads—Oiling and Lubricating—Tuner Cleaning—Demagnetizing—End-of-Tape Sensor Cleaning—Reel Spindles—Front Panel Controls—Final Inspection—Reassembly—Checkout—Maintenance Log—Remote Cleaning—Tape Troubles

Chapter 6   149

VCR First Aid

Tools and Safety—Dropped Deck—Fire Damage—Water Damage—Sand, Dirt, and Dust—Foreign Objects—Leaked Batteries

Chapter 7   160

Troubleshooting and Repairing Non-VCR Problems

Start at the TV—Program Reception—VCR Cables and Controls—Tracking Control—TV/VCR Switch—Remote Interference—Using a Known, Good Tape—Troubleshooting Checklist

Chapter 8   169

Troubleshooting Techniques and Procedures

The Essence of Troubleshooting—Troubleshooting Charts—Using the Troubleshooting Flowcharts in Chapter 9—What You Can and Can't Fix—What to Do if You Can't Fix It—Troubleshooting Techniques—Going Beyond the Flowcharts

Chapter 9   178

Troubleshooting VCR Malfunctions

Fix It Yourself—Service Politics—Maintenance Procedures and Special Adjustments—Using the Troubleshooting Flowcharts—Flowchart 1. VCR Does Not Turn On—Flowchart 2. VCR Turns On but Nothing Else—Flowchart 3. Cassette Will Not Load—Flowchart 4. Cassette Will Not Eject—Flowchart 5. VCR Will Not Thread Tape—Flowchart 6. VCR Will Not Play Tape—Flowchart 7. VCR Eats Tape—Flowchart 8. Fast-Forward or Rewind Won't Operate—Flowchart 9. Search Does Not Work Correctly—Flowchart 10. Tracking Control Has No Effect—Flowchart 11. VCR Does Not Respond to Some or All Front Panel Controls—Flowchart 12. Front Panel Indicators Not Functioning—Flowchart 13. Timer Does Not Operate Properly—Flowchart 14. Tuner Channels Don't Change on TV—Flowchart 15. Sound OK; Video Not OK—Flowchart 16. Video OK; Sound Not OK—Flowchart 17. Snowy Video; Poor Audio—Flowchart 18. Remote Control Does Not Operate or Function Properly—Flowchart 19. VCR Makes Unusual Mechanical Noises During Loading and Playback—Flowchart 20. You Receive an Electric Shock When You Touch the VCR—Flowchart 21. VCR Overheats—Miscellaneous VCR Difficulties

Chapter 10   265

VCR Reference Guide

Chapter 11   287

Camcorders

Disassembly—Head Cleaning—Dusting and General Cleaning—Parts Cleaning and Replacement—Camera Care

Appendix A   291
    Sources

Appendix B   296
    Further Reading
        Magazines—Books

Appendix C   299
    Maintenance Log

Appendix D   302
    Soldering Tips and Techniques
        Tools and Equipment—Basic Soldering—Replacing Components—A Good Solder Joint—
        Electrostatic Discharge—Iron Tip Maintenance and Cleanup

Appendix E   306
    Attaching an F-Connector
        What to Do—Tools List

Appendix F   309
    Cable and Component Signal Loss

Appendix G   311
    Television Frequency Spectrum

Appendix H   316
    VCR Specifications Charts

Glossary   323

Index   333

Acknowledgments

My father, Wally McComb, helped immensely with the research of this book; his efforts and insight are much appreciated. I also want to thank Southwest General Industries, of Carlsbad, California, for providing me with their service records and expertise, and to thank Roderick Woodcock, for loaning me a small portion of his encyclopedic knowledge of video. My wife Jennifer helped me through the rough spots and saw to it that I didn't miss my deadlines . . . much. Well, at least she tried. Finally, a warm thank you for the patience of my agent, Bill Gladstone, and to Brint Rutherford and the editors at TAB BOOKS Inc.

Introduction

Have you ever wished you could freeze time or wind back the clock to a happier moment? What if you could edit out the boring parts of your life? You can do all this and more. There's no strange science fiction gadgetry, haunting sorcery, or medieval magic involved. All you need is a video cassette recorder.

Video cassette recorders (VCRs) allow anyone to become a programming executive. You can record shows off the air for later viewing, tape important moments of time with the help of a video camera, and watch full-length movies in the comfort of your living room.

Home VCRs were introduced in 1976; consumers have bought over 12 million of them since. In the next two years, industry pundits believe, another 12 million will be sold. It is thought that by the end of the decade, over half of all American households—about 45 million in all—will have a video recorder. VCRs have gone past the point of being toys for Yuppies to that of essential appliances, as necessary for existence in this technological age as a refrigerator or oven.

But VCRs are machines, after all, and therein lies a problem. Machines need care, and they break down every now and then. The cost of sending a VCR to a video doctor is astronomical—as high as $100 for a simple belt change or adjustment. Cleaning and lubricating a VCR, a job that should be done every 12 to 18 months, costs upwards of $100. Owning a VCR can be an expensive proposition.

Though a VCR might be a bundle of high-tech circuitry, servo motors, and precision parts, keeping one in good health needn't be a complicated matter. In fact, taking care of a VCR isn't really any more difficult than taking care of your car. Maintaining and repairing a VCR takes but a small assortment of inexpensive tools and the right instructions. The tools can be purchased at most any hardware or electronics store, and the "right instructions" is a book called *Troubleshooting and Repairing VCRs*.

Troubleshooting and Repairing VCRs is designed expressly for the consumer. It offers step-by-step details on the care and feeding of home VCRs—from simple cleaning and lubricating of parts to troubleshooting power supply and logic circuitry problems. It is the first book to cover, in an in-depth but non-technical manner, the care and repair of all types of home video cassette recorders, including Beta, VHS, and 8mm, as well as the ever-popular camcorders (combination cameras and portable video cassette recorders).

The instructions on VCR maintenance and repair are simple and straightforward. Complex theory is kept to a minimum. Instead, there's a liberal blend of actual hands-on help, tips, and techniques—the kind of information that is needed the most.

This book is designed and written for the video hobbyist and electronics enthusiast. It is technical in nature, but it does not assume an intimate knowledge of video, audio, computer technology, electronics, or maintenance and repair techniques. You'll find plenty of introductory information, plus tips that will help you to not only enjoy your VCR investment, but save you time and money when (and if) repair time comes around.

About 90 percent of all VCR malfunctions are mechanical, so *Troubleshooting and Repairing VCRs* spends a lot of time discussing mechanical breakdown—how to avoid it and how to repair it. That doesn't mean the book lacks information on troubleshooting (and in some cases repairing) VCR circuitry. On the contrary, every major electronic subsystem of VCRs is fully covered, including power supplies, solenoid control, motors, and more.

All home VCRs work on the same basic principles, but specific maintenance and repair techniques can vary widely between models. *Troubleshooting and Repairing VCRs* contains specifications and repair notes for over four dozen different video recorders, including a handy cross-reference chart on models that are sold by many different firms but are made by only a handful.

Video cassette recorders are loaded with specialty parts like polished video heads, complex threading mechanisms, and proprietary surface mount integrated circuits. You can't get these things at the local Radio Shack, and most manufacturers don't sell replacement parts to consumers. Even if you could obtain the components, they require special alignment tools and test jigs to test and install properly.

All this means that in the event your VCR has a truly serious problem, the home-based technician can do little to affect repairs. Unless you have the specific knowledge of servicing your particular brand of player, an oscilloscope to diagnose waveforms, all the specialty tools on hand, and a service manual, it is better having serious ailments serviced at a repair center. You are free to attempt larger scale repairs, of course, but they are beyond the scope of this book.

Fortunately, malfunctions in the crucial components of VCRs are rather rare, even with the inexpensive models. Most problems are caused by such things as dirty switch contacts, broken wires, dirty video heads, old and worn rubber belts and rollers, and damaged tapes. In fact, these represent the greatest percentage of service calls to repair centers. You can fix these faults yourself, and with a minimum of tools and time. This book shows you how.

You can greatly minimize repairs—whether done by you or someone else—by keeping your VCR in tip-top shape. This book presents an easy-to-follow preventative maintenance schedule that you can use to keep your deck working at its fullest. Even if you can't repair the VCR yourself, this book serves another important purpose: it helps you to be well informed about the possible causes of video cassette recorder problems. You will be better able to articulate your deck's illness to the repair technician, and by specifically stating what is wrong, you have a greater chance that the problem will be fixed correctly the first time, and at a lower cost.

You will also be in a better position to spot unscrupulous repair tactics, like charging for parts that were never replaced, or labor above and beyond what had to be done to service a component. Most service centers and repair technicians are honest and fair, but there are exceptions. Be on the lookout for them!

Here is a quick rundown of what you'll find in each chapter of *Troubleshooting and Repairing VCRs.*

Chapter 1—Introduction to Video Cassette Recorders. The basics of video and video cassette operation, including all the latest technological advances: HQ, Super-VHS, digital effects, and hi-fi audio.

Chapter 2—How VCRs Work. An in-depth look at how VCRs record and play back tapes, differences between VHS, 8mm, and Beta, and operating controls.

Chapter 3—The VCR Environment. Setting up a VCR in your home, avoiding problems in installation, how to properly hook up a VCR to a TV, hi-fi, antenna, and other components to enhance the picture and sound.

Chapter 4—Tools and Supplies for VCR Maintenance. Personal safety, basic tools for disassembly, volt-ohm meters and their use, logic probes, scopes and frequency counters, cleaning supplies, lubricating oils and greases, building an infrared light sensor.

Chapter 5—General Cleaning and Preventative Maintenance. PM schedules and procedures for cleaning the video heads, interior cleaning, lubrication, checkout, tape care and repair, remote control maintenance.

Chapter 6—VCR First Aid. Emergency procedures for minimizing or eliminating damage caused by fire, water, dirt, sand, and other contaminants.

Chapter 7—Troubleshooting and Repairing Non-VCR Problems. How to identify and correct video and audio problems that are not caused by the VCR, including bad tapes, improper hookup, and a maladjusted or malfunctioning TV.

Chapter 8—Troubleshooting Techniques and Procedures. A logical approach to VCR repair, how to analyze problems, how to use the flowcharts in Chapter 9, and proper troubleshooting etiquette.

Chapter 9—Troubleshooting Charts for VCR Malfunctions. Flowcharts for a variety of common ailments, including decks that don't turn on, don't play tapes, only work for a few moments then stop, ignore front panel controls, and more.

Chapter 10—VCR Reference Guide. Specifications and features for popular VCRs.

Chapter 11—Camcorders. Routine maintenance and repairs.

Appendix A—Sources. Names and addresses of VCR manufacturers.

Appendix B—Further Reading. Suggested reading to broaden your knowledge of video and VCRs.

Appendix C—Maintenance Log. A four-page form for use when cleaning and servicing your VCR. Make copies and keep with the deck's instruction manual.

Appendix D—Soldering Tips and Techniques. Helpful instructions on how to solder and desolder components and wires.

Appendix E—Attaching an F-Connector. How to attach an F-connector to a 75-ohm coaxial cable.

Appendix F—Cable Loss Chart. Average signal loss through standard coaxial cable types.

Appendix G—Television Frequency Spectrum. A chart of the TV frequency spectrum, and how it is broken down into channels.

Appendix H—VCR Specifications Chart. A cross-reference chart of over 300 early and late model VCRs and camcorders.

1

Introduction to Video Cassette Recorders

To appreciate the advances that have been made in video, let's suppose that you could turn the clock back in time. Imagine that you have found a time machine in your basement. You step inside the machine and go back to the day May 20, 1966. You arrive at 7:30 in the evening, in time to watch the evening's fare of television.

One of the first things you might notice is that although the channels are full of programming, you're limited to what you can view and when. Tune into ABC and you get the "Green Hornet." No, you're not a Bruce Lee fan, so you spin the dial and find "The Wild, Wild West" on CBS, and "Tarzan" on NBC. Still not right. Your only other choice is to watch a 1940's war movie, which seem to be in steady supply on many of the local TV stations. After the evening is over, you realize that you enjoyed living in the "present" much better because you have control over your television viewing schedule.

You zap yourself back to the future where you are not limited to the programming tastes of television network executives. Thanks to video cassette recorders (VCRs), you can be your own programming executive—you set your own prime-time schedule to match your needs and viewing tastes.

VCRs come in many different shapes and sizes. Let's take a close look at VCRs, and what the various features and functions do. This book investigates the three formats of VCRs and the kinds of features found in all models. This chapter gives an overview of video cassette recorders. Chapter 2 details how video recorders work and provides a broad overview of the science and technology found in the typical VCR.

INTRODUCTION TO VIDEO CASSETTE RECORDERS

VCR FORMATS

All VCRs work about the same. They record sound and picture information on magnetic tape, then play that information back on your TV. VCRs might be based on the same technology, but inside they are worlds apart. The first major differences between models is that there are three popular formats, VHS, Beta, and 8mm. An example of a VHS deck is shown in Fig. 1-1.

Inside the Three Formats

Beta is the brainchild of the electronics giant Sony. Though it was not the first consumer video tape format to come out, it was the first to catch on. Sony's first Beta deck was introduced in 1975 and became widely available in 1976. Since then, Sony and a handful of other manufacturers have come out with several dozen models of Beta VCRs. Beta tapes are ½-inch wide and are enclosed in a plastic cassette. The cassette protects the delicate tape from dust, dirt, and other contaminants.

A few years later, another Japanese electronics firm, JVC, introduced a rival VCR tape format. They dubbed the newcomer VHS—the acronym stands for Video Home System. Like Beta, VHS tapes are ½-inch wide and are enclosed in a plastic

Fig. 1-1. One of several hundred models of VCRs to come out since 1976.

Fig. 1-2. Beta (foreground) and VHS cassettes.

cassette. However, the cassette is physically larger, as shown in Fig. 1-2, and the signals are recorded on the tape in a different fashion than Beta. As a result, VHS tapes won't play on Beta decks and vice versa.

Through the years, the VHS format has achieved the most success, and as of this writing, Sony is the only remaining maker of Beta decks. Even Sony has greatly curtailed manufacture and marketing of Beta products; there are only a small handful of Beta decks available, and most of these are high-end, semi-professional models.

In 1982, Sony and 127 other manufacturers from around the world joined forces to create a new VCR format using tape that is eight millimeters wide. The new format is called 8mm, and is the most popular with camcorders (combination cameras and recorders). Tape cassettes for 8mm decks are only slightly larger than audio cassettes (see Fig. 1-3), so 8mm VCRs can be made very small.

Playback Speeds

The first VCRs in all three formats recorded and played back at one speed only. This allowed a maximum of one or two hours or so of recording time on a single

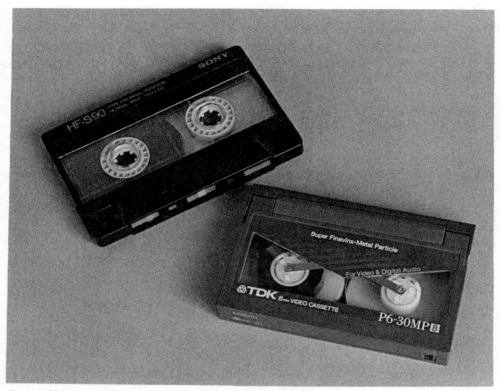

Fig. 1-3. 8mm cassettes are roughly the same size as audio cassette tapes.

cassette. Rather than try to pack four or six hours worth of tape in a cassette—which is difficult at best—VCR manufacturers decided to enable alteration of the speed of the tape traveling through the deck.

Most VHS decks have three record/playback speeds: SP, LP, and EP (also called SLP on older models). Home Beta decks can record and playback in two speeds, BII and BIII (the fast BI record speed is reserved for industrial and special semi-professional decks only, but a number of home Beta decks can play back BI tapes). The 8mm decks have two speeds, SP and EP. Table 1-1 shows how the recording times measure up at the various speeds, based on commonly available tapes.

Slowing down the tape decreases the quality of the picture and sound. The best quality is obtained by recording at the fast speed. The reason is made clear in Chapter 2, "How VCRs Work."

VCR TYPES

Most all VCRs are the tabletop variety; they are designed for in-home use and plug into the power outlet for electricity. But there are several other "sub-types" available.

Table 1-1. Recording Speeds.

VHS	SP	LP	EP (SLP)
T-60	1 hr	2 hrs	3 hrs
T-120	2 hrs	4 hrs	6 hrs
T-160	2 hrs 40 min	5 hrs 20 min	8 hrs
Beta	**BI**	**BII**	**BIII**
L-125	15 min	30 min	45 min
L-250	30 min	1 hr	1 hr 30 min
L-500	1 hr	2 hrs	3 hrs
L-750	1 hr 30 min	3 hrs	4 hrs 30 min
L-830	N/A	3 hrs 30 min	5 hrs
8mm	**SP**	**LP**	
MP-30	30 min	1 hr	
MP-60	1 hr	2 hrs	

Camcorders

Up until a few years ago, portable VCRs were popular. These operated off battery power and, when connected to a video camera, allowed you to shoot your own tapes anywhere. Portables are no longer manufactured; they have been replaced by a combination VCR and camera in one box, called the camcorder (see Fig. 1-4). Camcorders are especially useful if you do a lot of away-from-home shooting, because you're not encumbered by a bulky VCR, camera, and cables. Everything is self-contained. Some camcorders are record-only (you must use a tabletop VCR to watch the tape), but the vast majority have both record and playback capability.

If you have a video camera, you can use it with either a portable or tabletop model. The majority of cameras are made for portable VCRs and attach to the deck via a specialized 10- or 14-pin connector (10-pin is standard; the 14-pin type is found on some older Beta decks). Adapters are available to match the camera to the type and model of VCR. More information on camcorders is contained in Chapter 11.

Video Cassette Players

Video cassette players, or VCPs, lack a recording capability—they play back tapes only and lack the means to record. This means VCPs dispense with a lot of extraneous electronics, as well as the programmable record timer. As a result, VCPs are a little less expensive than a full-fledged recording VCR. However, they require the same level of mechanical maintenance as full-fledged recording VCRs.

VIDEO AND AUDIO HEADS

VCRs record and play back audio and video signals using magnetic heads. At a minimum, all VCRs use the following magnetic heads:

✦ A stationary full erase head.
✦ A stationary audio head.

Fig. 1-4. A camcorder—combination video camera and recorder in one unit.

◆ Rotating video heads.
◆ A stationary control track head.

The way these heads record information on the tape is similar for the three formats, though the exact track widths and specifications vary. The recording formats for VHS, Beta, and 8mm are shown in Fig. 1-5. The relative location of the heads in the types of VCRs is shown in Fig. 1-6.

Full-Erase Head

The full-erase head erases the entire tape and is on only during recording. The erase head "blanks out" any previous recordings that might be on the tape. In certain VCRs, a secondary audio erase head can be used. This head erases just the audio track. For more details, see the section below on "Additional Heads".

Audio Head

In the record mode, the tape moves past the audio head, where the head impresses the sound signal on a thin track on the edge of the tape. In the playback mode, the head picks up the signal recorded on the tape.

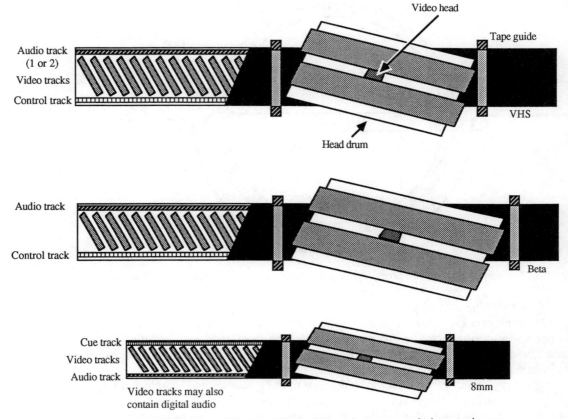

Video head

Tape guide

Audio track
(1 or 2)

Video tracks

Control track

Head drum

VHS

Audio track

Control track

Beta

Cue track

Video tracks

Audio track

Video tracks may also
contain digital audio

8mm

Fig. 1-5. The three home video formats and how the information tracks are recorded on each one.

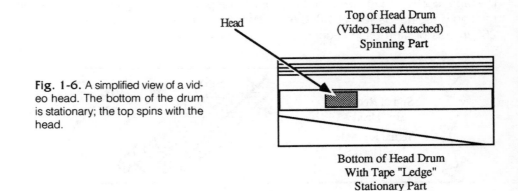

Head

Top of Head Drum
(Video Head Attached)
Spinning Part

Fig. 1-6. A simplified view of a video head. The bottom of the drum is stationary; the top spins with the head.

Bottom of Head Drum
With Tape "Ledge"
Stationary Part

Side View

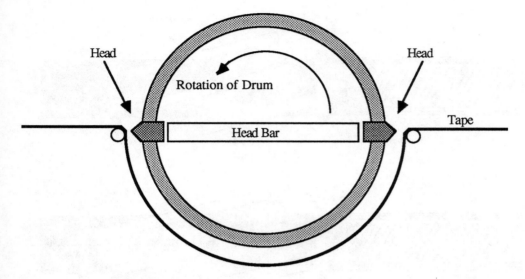

Fig. 1-7. Top view of a video head drum, showing both heads and the 180-degree path of the tape.

Video Heads

Video information is much more complex than audio information. More tape area is needed to record a picture. To pack a video picture on a ½-inch or 8mm tape using conventional recording techniques, the tape must be shuttled through the deck at speeds approaching 20 feet per second.

Obviously this is impractical, so VCRs use rotating heads, as illustrated in Fig. 1-7, to record and play back video. The heads are mounted within a polished metal cylinder, called the *head drum* (other names are used as well, including *scanner, head cylinder,* and *headwheel*).

The heads are at an angle, so they record a series of long, diagonal tracks on the tape, shown in Fig. 1-8. The tape might creep slowly through the VCR, but the rotating video heads spin at 1,800 revolutions per minute, effectively covering about 250 inches of tape each second.

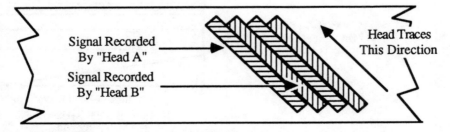

Fig. 1-8. The heads on the head drum record adjacent tracks on the tape.

Table 1-2. Writing Speed.

Format and Speed	Writing Speed	Linear Tape Speed
Beta		
B-I	274.6 ips (6975mm/s)	1.57 ips (40mm/s)
B-II	275.4 ips (6995mm/s)	0.78 ips (20mm/s)
B-III	275.6 ips (7002mm/s)	0.52 ips (13.3mm/s)
VHS		
SP	228.5 ips (5804mm/s)	1.31 ips (33.35mm/s)
LP	229.1 ips (5820mm/s)	0.66 ips (16.7mm/s)
EP	229.3 ips (5826mm/s)	0.44 ips (11.12mm/s)
8mm		
SP	147.7 ips (3751mm/s)	0.56 ips (14.3mm/s)
LP	148 ips (3758mm/s)	0.28 ips (7.2mm/s)

Note: Writing speed increased slightly at slower linear tape traveling speeds

Table 1-2 shows the relative speeds of the tape traveling through a VHS VCR versus ''writing speed,'' which is the amount of tape covered by the video heads in one second.

Number of Video Heads. All VCRs (except the Betamovie portable camcorders) use a minimum of two video heads mounted 180 degrees apart. Only one head touches the tape at any time. The more advanced VCRs incorporate even more video heads for improved playback at all speeds plus high-quality special effects. See Chapter 2, ''How VCRs Work,'' for a more detailed discussion of what each of the video heads do.

Control Track Head

During recording, the control track records a series of 30-hertz pulses. These pulses are used to synchronize the video heads during playback so that they pass directly over the tracks that were previously recorded. Without the control track, the video heads might not scan directly over the video tracks, and the picture would be garbled. The control track serves the same general purpose as sprocket holes in movie film. The sprocket holes help align each frame so that you see a steady picture on the screen.

In most all VCRs, the control track head is mounted in the same unit as the audio head.

Additional Heads

Some VCR models incorporate additional heads. Their functions are discussed below.

Stereo Audio Heads. Monophonic VCRs use one audio head to record and play back sound. Stereo VHS decks use a dual audio head for recording the right

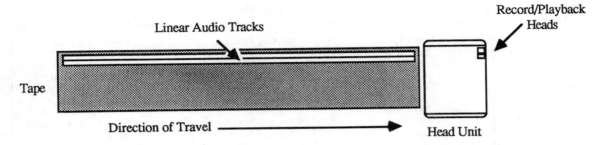

Fig. 1-9. A linear stereo head records the right and left soundtrack separately at the top of the tape.

and left tracks separately. Audio recording with two audio heads is called linear stereo. The tracks are placed along side one another near the edge of the tape (see Fig. 1-9) and occupy the same space as the standard mono audio track (monophonic decks pick up both the right and left channels together). Noise reduction circuitry (such as Dolby) is often used with linear stereo decks to improve the sound quality.

On a number of linear stereo VCRs, one or both of the channels can be independently erased and/or activated for audio dubbing. The audio erase head is built into the audio head, as depicted in Fig. 1-10. With audio dubbing, you record only the audio segment, but leave the video and control track intact. Most linear stereo decks let you record on either the left or right stereo channel, leaving the other channel untouched.

Beta VCRs record stereo audio mixed with the picture information, as explained below.

Rotating Audio Heads. VHS hi-fi decks—those that record stereo sound in high-fidelity—have two additional heads mounted with the rotating video heads. These heads record audio information in an FM carrier signal. Even though the audio and video tracks are intermixed, they do not interfere with one another because the angles

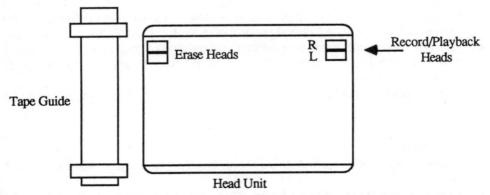

Fig. 1-10. Close-up view of the control/audio head unit. The one unit contains the control and audio heads, and can also contain one or more audio erase heads.

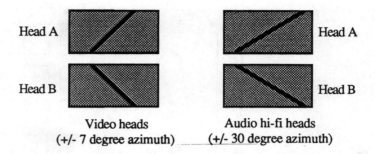

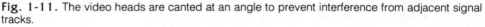

Fig. 1-11. The video heads are canted at an angle to prevent interference from adjacent signal tracks.

(or azimuth) of the heads are different, as depicted in Fig. 1-11. Magnetic heads do not pick up information that is recorded on substantially different angles. The video heads themselves also record at different azimuth angles, as discussed more fully in Chapter 2, ''How VCRs Work.''

Beta stereo hi-fi decks record picture and high fidelity sound using the machine's standard complement of video heads. Both the sound and picture are combined in one signal and placed on the tape together.

Flying Erase Head. The main full erase head expunges the contents of the entire tape prior to recording. The 8mm tabletop VCRs, as well as many 8mm and VHS camcorders, use a flying erase head that is mounted with the video heads. The flying erase head erases just the video tracks, leaving the control and audio tracks alone. The flying erase head provides for better, cleaner edits.

TUNING

VCRs are designed so that they can record programming from a TV broadcast antenna or from cable. Like your TV, VCRs incorporate tuners, so you can select the channel you want to record.

Older VCRs used mechanical *detent* tuners, the same as the old-style TV sets. Besides not being able to change the channel under program or remote control, mechanical detent tuners usually require periodic cleaning and can go out of alignment . Dirty contacts in the tuner can cause a variety of problems, such as loss of signal quality and noisy reception. Later chapters detail how to clean tuner contacts.

All current VCRs come with electronic pushbutton tuning, as shown in Fig. 1-12. You have access to VHF channels 2 through 13, as well as UHF channels 14 through 83. Most decks let you preset 12 or more channels so you don't have to flip through every channel to find the station you want. The better models have random access tuning, letting you dial in the exact station on a keypad.

The pushbuttons used in an electronic tuner might also need occasional topical cleaning as covered in future chapters, but they aren't as susceptible to the same trouble that plagues mechanical tuners.

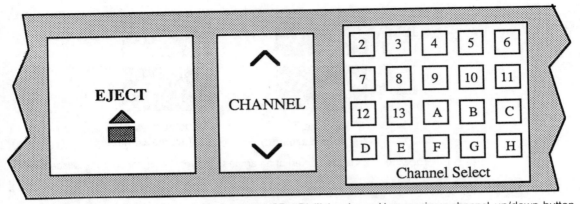

Fig. 1-12. A typical electronic tuner on the front of a VCR. "Dial" the channel by pressing a channel up/down button or a button in a keypad.

Cable-Ready

Cable systems that carry more than 12 channels usually broadcast the extra channels on special cable frequencies that normal TVs and VCRs can't receive. However, a cable-ready VCR (or TV for that matter), *can* tune to these channels. The number of accessible channels varies among VCRs but is usually from 102 to over 140. Frequency allotments for regular and cable channels is provided in Appendix G.

Having a cable-ready VCR means you don't need to use the cable company's converter box to watch the non-scrambled channels. Bear in mind that a cable-ready VCR or TV *will not decode a scrambled picture.* You'll still need a cable TV decoder box for this.

Programmable VCRs

One of the benefits of owning a VCR is being able to record programs while you're away from home. Nearly all VCRs have at least a rudimentary timer built into them (the timer was a separate component on the first VCRs). You select the channel you want, then set the timer to start recording at a particular hour (and sometimes day). You can program the timer to record for as little as a minute, or as long as you want, up to the maximum length of the tape. This type of recording is called *time shifting.*

More elaborate VCRs, especially those with electronic tuners, allow you to record several different programs, at different times of the day, on different days of the week, and on other channels. You could, for instance, record the news from 8:00 to 8:30 in the morning, Perry Mason at noon, and cartoons for the kids from 2:00 to 3:30.

VCR timers are rated by the number of days and events they can be programmed for. VCRs typically have timers that allow recording of up to seven or 14 programs over a seven day period. The only limitation is that you can't capture more programming than the tape will hold.

OPERATING CONTROLS

The earliest VCRs used clunky mechanical controls for their various operating modes, such as PLAY, REWIND, and FAST FORWARD. You pushed a button, that turned a lever, that rotated a crank, that flipped a ratchet, that pushed down a spring, that moved a belt, and so forth. Fully mechanical VCRs are no longer made, but if you own one, you will need to service it from time to time to keep the springs, ratchets, belts, and other components operating smoothly. Chapter 5 deals with these maintenance procedures.

Now, all VCRs use touch-sensitive electronic or electromagnetic controls. When you push a button, an electrical signal is sent to a motor or solenoid, bypassing all the intermediate mechanical linkages. Note that the machine is less complex, mechanically, but a simple break in the signal lines from the switch to the actuating motor or solenoid could mean an inoperative deck.

Remote Control

Remote controls come in two forms: wired and infrared. Wired remotes are tethered to the VCR with a cable and are usually limited in function. You can stop and start the VCR but little else. Infrared remotes, which work by sending pulses of invisible light to your VCR, incorporate all or nearly all of the functions available at the deck itself.

Many of the early VCRs (pre 1983/1984) used wired remotes, but now VCRs use infrared wireless units. Maintenance and service procedures for both types of remotes are provided in Chapters 5 and 9.

BASIC OPERATION

All VCRs have the features discussed in the following sections.

Tape Loading

Tape loading is via manual or automatic means. Manual tape loading, found on all top-loading VCRs (no longer made except for camcorders) involves inserting a tape into the loading mechanism and pushing the mechanism down into play position. Automatic tape loading, found on all front-loading VCRs, involves inserting a cassette into a loading door. The VCR grabs the tape and pulls it inside the machine. Figure 1-13 shows a top-load VCR and Fig. 1-14 shows a front-load VCR.

Function Controls

The basic operator controls are PLAY, RECORD, FAST-FORWARD, REWIND, and PAUSE. The PLAY button plays a previously recorded tape; the RECORD buttons make a new tape. The FAST-FORWARD and REWIND buttons shuttle tape back and forth within the cassette, and the PAUSE button temporarily stops the VCR but does not fully disengage recording or playback.

Fig. 1-13. An older model top-load VCR.

Fig. 1-14. A front-load VCR (with remote control).

Input/Output Terminals

The input and output terminals serve as gateways in and out of the VCR. The input terminals provide a signal for the VCR to record. Conversely, the output terminals serve to route the playback signals from the VCR to a television set.

Two types of terminals are used: rf and baseband video/audio. Rf signals combine both audio and video on a modulated carrier and are broadcast on a television channel, usually channel 3. Baseband signals consist of only raw audio and video, and are used to connect to a monitor or other VCR. For more details on baseband and rf signals, see Chapter 2.

THE ENHANCED VCR

There are plenty of special features available on VCRs. Here's a run-down of the enhancements you might see on decks in just about every price range.

Special Video Effects

If you're a TV sports nut or have ever even watched a football game on TV, you've seen special video effects at work. Many home VCRs are capable of the special effects used in sports broadcasts, including slow motion and freeze-frame.

Slow motion lets you slow things down so you can better analyze fast action, while freeze-frame lets you stop the action altogether. For freeze-frame, you find the frame you want, and press the pause (or still) button on the VCR. Freeze-frame is available on most models, but slow motion is a feature usually found on higher-end models only. For slow motion, you can usually vary the speed from very slow to half the normal speed.

For relatively noise-free pictures—in freeze-frame or slow motion modes—VCRs have one or more special-effects video heads, as discussed in Chapter 2, "How VCRs Work." In addition, the latest VCRs incorporate digital circuitry to get rid of the tell-tale "noise" you see when freezing a frame.

Because of the way special-effects heads work, freeze-frame and slow motion might not be available at all speeds on some decks. Typically, special video effects are not available at the LP speed on a VHS machine, or the BII speed on some Beta decks.

Many VCRs have the opposite of slow motion, called fast scan. With fast scan (also called *cue* or *review*), the action speeds by quickly, so you can find a particular spot on the tape without having to go back and forth between play and fast forward or rewind. The fast scan speed varies depending on the machine and the speed that the tape was originally recorded. It's typically from three to ten times faster than normal play.

Stereo Ready and Stereo Adaptable

Being able to record and play back tapes in stereo doesn't mean the VCR can receive television broadcasts in stereo. That requires a deck that's either *stereo ready* or *stereo adaptable*.

➟ **Stereo ready** means that the VCR has an MTS (multichannel TV sound) tuner built into it, so that without extra circuitry the deck can receive and process programs broadcast to you in stereo.

➟ **Stereo adaptable** means that the VCR can be connected to a decoder that includes the circuitry necessary to receive and process the stereo sound. Without the box, the TV processes monaural sound only. A VCR that is stereo adaptable has an MPX, or *multiplex,* jack on it, for connection to the decoder.

Audio and Video Dubbing

When you place a VCR in the record mode, it tapes both the sound and picture at the same time. Some top-of-the-line models let you record only audio or video, leaving existing sound or picture signals intact.

Let's say you've videotaped your collection of Super 8mm home movies. Unless they're sound movies, you won't have a soundtrack to accompany the picture, but that doesn't mean you have to sit through silent movies whenever you watch the tape. An audio dub facility allows you to record sound *after* the picture has been placed on the tape, as illustrated in Fig. 1-15.

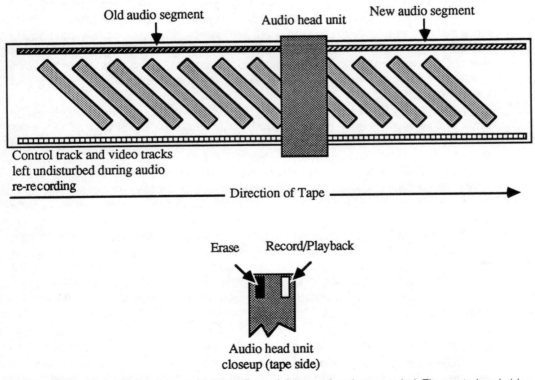

Fig. 1-15. During audio dubbing, only the audio track is erased and re-recorded. The control and video tracks are unaffected.

If the VCR is stereo, you can usually record the left or right channels independently (see the discussion of audio heads previously in this chapter). This independence lets you add sound during one recording session (on the right channel, for example) and narration during another recording session (in this case, the left channel). Video dubbing records new picture over old, without harming the audio portion.

Super Beta

In 1984, Sony introduced a new signal processing scheme that squeezed out extra picture quality from Beta decks. The new Super Beta VCRs can deliver a signal up to 20 percent sharper than standard Beta models. The actual increase in picture quality you enjoy is also determined by the quality of tape you're using, the source program, and other variables. If you have a low performance TV, for example, you won't likely see much of an improvement at all.

Tapes recorded on a Super Beta deck can be played back on a standard deck, but without the improvement in picture quality. Likewise, a Super Beta deck can't deliver higher quality pictures when playing back a tape recorded on a conventional Beta machine.

HQ

Many of the latest VHS VCRs use circuitry to artificially enhance the quality. The circuits are collectively known as *HQ,* which stands for high quality. Here is rundown of the HQ circuits and what they do.

- **White Clip Level Enhancement.** All VCRs cut off scenes that exceed a certain brightness. The cut off point is known as the white clip level. In HQ VCRs, the white clip is extended by another 20 percent, thus increasing the apparent sharpness of details and outlines.
- **Brightness Noise Reduction.** Brightness noise looks like grains of sand sprinkled over the picture. The more noise, the less sharp the picture. Noise is random, and the VCR never knows where it will occur on the screen. The noise reduction circuits duplicate the video information for the whole screen, then average the result, so the picture is apparently sharper.
- **Color Noise Reduction.** Color noise shows up as streaking, and is reduced in a similar fashion as brightness noise.
- **Detail Enhancer.** Detail enhancer strengthens the amplitude of high frequency signals (these carry the ''sharpness'' information).

Not all makers of VHS VCRs have adopted HQ and not all makes employ every one of the four separate circuits described above. Often, only two or three of the circuits will be used, such as the brightness noise reducer and detail enhancer or the white clip. A few VCRs let you vary the amount of noise reduction or detail enhancement or turn the HQ circuits off altogether.

INTRODUCTION TO VIDEO CASSETTE RECORDERS

Super VHS

Super VHS is a sub-format of VHS where the picture resolution is increased by separating the brightness and color components and by widening the range of the brightness signal. The unusual signal scheme of Super VHS make tapes recorded on a Super VHS unwatchable on a regular VHS deck. However, all Super VHS machines can play standard VHS tapes, and you can manually switch the Super deck to record in normal VHS format. This assures you of being able to watch pre-recorded movies and share tapes with friends and relatives who do not own Super VHS. There's more about this interesting sub-format in the text chapter.

CTL

CTL stands for "control," and is a coding method using the control track of VHS video cassette recorders to enable you to quickly find a certain segment on the tape. The CTL system, found on only some high-end decks, provides two types of coding: *index* and *numeric* address. With index coding, you identify the start or some other portion of a program with "markers." Indexed points are sensed during playback and fast-scan (forward or reverse). Depending on the model of VCR, the index codes can be added during initial recording of the program or after.

The VHS address search system allows you to place a multiple-digit code onto the tape. The address code can be written either manually (by you) or automatically (by the VCR or a personal computer) during recording or playback. You might, for example, use the address code system to number the programs as they appear on a tape, to memorize the play or record time, or even to jot down the time and date of the recording.

Digital Effects VCRs

Digital effects VCRs use computer circuitry to produce blemish-free special effects during playback. With all VCRs, pausing the tape during playback yields "noise bars" or lines at the top and bottom of the screen.

Digital circuits freeze an instant of video by capturing the image into computer memory, then relaying the image stored in memory to the TV set. The tape itself can keep moving within the VCR; the image on the screen is frozen in silicon and doesn't change until you hit the FREEZE or PAUSE button again. More details on digital effects VCRs can be found in the next chapter.

2

How VCRs Work

Understanding the way VCRs work is an integral part of knowing how to fix them when they break. Although routine maintenance procedures and home repair do not require a scientific explanation of such things as how the heads in a VCR magnetically record information on tape, it is helpful to know the basics of VCR operation.

This chapter details the technical side of the inside workings of VCRs, with full discussions on the tape loading and threading schemes used by Beta and VHS decks, how the tape is transported within the machine, how the number of video heads determines picture quality at all playback speeds, and more.

In many ways, VCRs are like television sets without the picture tube. To fully understand how VCRs play back and record picture and sound, you need to know how television works. Hence, in this chapter are details on the science of television— how electronic images are beamed through the air and received and processed by your TV and VCR. Understand basic TV first, then graduate to the operation of VCRs.

TELEVISION HISTORY 101

Though it seems as if television has been around for a long time, it's a relatively new science, younger than rocketry, internal medicine, and nuclear physics. In fact, some of the people that helped develop the first commercial TV sets and erect the first TV broadcast antennas are still living today.

The first electronic transmission of a picture was believed to be made by a Scotsman, John Logie Baird, in the cold month of February, 1924. His subject was a Maltese Cross, transmitted through the air by the magic of television (also called "Televisor" or "Radiovision" in those days) the entire distance of ten feet.

To say that Baird's contraption was crude is an understatement. His Televisor was made from a cardboard scanning disk, some darning needles, a few discarded electric motors, piano wire, glue, and other assorted odds and ends. The picture reproduced by the original Baird Televisor was extremely difficult to see—a shadow, at best.

Until about 1928, other amateur radiovision enthusiasts toyed around with Baird's basic design, whittling long hours in the basement transmitting Maltese Crosses, model airplanes, flags, and anything else that would stay still long enough under the intense light required to produce an image. (As an interesting aside, Baird's lighting for his 1924 Maltese Cross transmission required 2,000 volts of power, produced by a room-full of batteries. So much heat was generated by the lighting equipment that Baird eventually burned his laboratory down.)

Baird's electro-mechanical approach to television led the way to future developments of transmitting and receiving pictures. The nature of the Baird Televisor, however, limited the clarity and stability of images. Most of the sets made and sold in those days required the viewer to peer through a glass lens to watch the screen, which was seldom over seven by ten inches in size. What's more, the majority of screens had an annoying orange glow that often marred reception and irritated the eyes.

In the early 1930's, Vladimir Zworykin developed a device known as the iconoscope camera. About the same time, Philo T. Farnsworth was putting the finishing touches on the image dissector tube, a gizmo that proved to be the forerunner to the modern cathode ray tube or CRT—the everyday picture tube. These two devices paved the way to the TV sets we know and cherish today.

The first commercially available modern-day cathode ray tube televisions were available in about 1936. Tens of thousands of these sets were sold throughout the United States and Great Britain, even though there were no regular television broadcasts until 1939, when RCA started what was to become the first American television network, NBC. Incidentally, the first true network transmission was in early 1940, between NBC's sister stations WNBT in New York City (now WNBC-TV) and WRGB in Schenectady.

World War II greatly hampered the development of television, and during 1941 and 1945, no television sets were commercially produced (engineers were too busy perfecting the radar, which, interestingly enough, contributed significantly to the development of conventional TV). But after the war, the television industry boomed. Television sets were selling like hotcakes, even though they cost an average of $650 (based on average wage earnings, that's equivalent to about $4,000 today).

Progress took a giant step in 1948 and 1949 when the four American networks, NBC, CBS, ABC, and Dumont, introduced quality, "class-act" programming, which at the time included Kraft Television Theatre, Howdy Doody, and The Texaco Star Theatre with Milton Berle. These famous stars of the stage and radio made people want to own a television set.

Since the late 1940's, television technology has continued to improve and mature. Color came on December 17, 1953 when the FCC approved RCA's all-electronic system, thus ending a bitter, four-year bout between CBS and RCA over color transmission standards. Television images beamed via space satellite caught the public's fancy in July of 1962 when Telstar 1 relayed images of AT&T chairman Frederick R. Kappell from the U.S. to Great Britain. Pay-TV came and went several times in the 1950's, 60's and 70's; modern-day professional commercial videotape machines were demonstrated in 1956 by Ampex; and home video recorders appeared on retail shelves by early 1976.

WHAT IS TELEVISION?

In an old "Twilight Zone" episode, a gun-toting desperado from the 19th Century is accidentally time-ported to present day. While walking through the busy streets of New York, he becomes unsettled by cars and bothered by the tall skyscrapers. But he totally falls apart when he sees a television set. There, displayed on the TV, is a classic Western: The good guy tells the bad guy to lay down his gun—or else. What does the desperado do? Why, he draws his gun and blasts the TV to tiny bits. Though the picture displayed on the TV was only in black and white, it looked like the real thing to the OK Coral reject. He saw a sharp, stable picture, both bright and vivid.

What the desperado didn't know is that what he thought was a real picture was actually a beam of light sweeping rapidly, line by line, over the face of a glass tube. The electron beam in a TV tube moves so rapidly and with such precision that the eye doesn't detect it, but fuses it into a steady, complete picture (the "persistence of vision" that Edison relied on when he invented the motion picture camera and projector also work with television images). If you were to slow down the beam, however, you'd see it start from the upper left corner, sweep over the face of the tube from left to right and top to bottom, as depicted in Fig. 2-1. This process would repeat over and over again.

Fig. 2-1. The electron beam of a TV tube scanning the face of the screen.

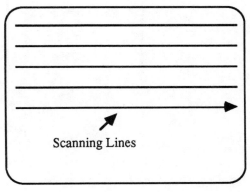

Scanning Lines

Breakdown of TV Signals

The TV tube in the television set is controlled by the incoming TV signal. This signal, after it is tuned to a specific channel and the audio is separated and sent to the audio amplifier, consists of many different sub-signals placed on a 4.2 MHz-wide band. This band is shown in Fig. 2-2. The component TV signals are:

- ◆ Picture brightness information, called luminance.
- ◆ Color subcarrier signal, called chrominance.
- ◆ 60-Hz vertical sync signal.
- ◆ 15.7-kHz horizontal sync signal.

Note that when the signal is transmitted through the airwaves (or sent through a cable), the picture is amplitude modulated (AM) and the sound is frequency modulated (FM). The audio carrier is placed at 4.5 MHz, which is just outside the luminance envelope.

The luminance (Y) signal comprises the bulk of the television transmission. The image information conveyed in the TV picture is vast, so it requires a large bandwidth, essentially from 100 Hz to 4.2 MHz. Note that the frequency of the information conveyed in the luminance signal determines the detail of the picture. A low-frequency signal creates a large, indistinguishable blur on the screen, while a high frequency signal creates a well-defined, pin-point shape.

The color (C) subcarrier signal, called chroma (also chrominance or "burst") is present only during color transmission. The science of color television transmission is involved and is beyond the scope of this book. But it is sufficient to say that the chroma subcarrier contains the information for both the color and color intensity for every instant of the transmission. The chroma subcarrier is phase-dependent; when the phase of the signal changes, the color changes. The instantaneous amplitude of the chroma signal determines the intensity or saturation of the color.

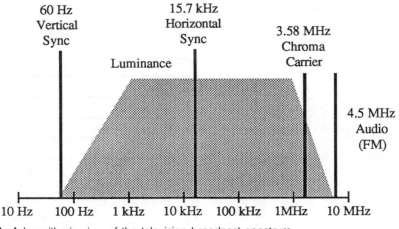

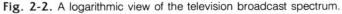

Fig. 2-2. A logarithmic view of the television broadcast spectrum.

Bear in mind that there is more to color TV than this, and if you are interested in learning more, refer to Appendix A, "Sources," for a list of books that discuss the topic in more detail.

The 60-Hz vertical synchronization is used by the television picture tube to direct the beam from the bottom of the screen back up to the top. This process is repeated 60 times per second, so the vertical sync rate is 60 Hz.

The 15.7 kHz horizontal synchronization is used by the television picture tube to direct the beam from left to right over the inside face of the tube. This process is repeated 15,734 times per second, so the exact horizontal sync rate is 15.735 kHz. For convenience, we'll refer to the horizontal sync rate as 15.7 kHz.

Note that all of these signals do not occur at the same time. In fact, they are precisely timed so that the TV set can adequately react to them, and in the proper order. Figure 2-3 shows a brief moment of television transmission. It illustrates the stream of signal transmitted to the television set and how it is broken down into the components we have discussed.

Inside the Picture Tube

If you dissect a picture tube, you can more fully understand the role that these signals play (for the time being, we will ignore the chroma signal). At the base of the TV tube is an electron gun. It emits a steady stream of electrons that strike the inside face of the tube, which is coated with a phosphor. When the electrons impinge on the phosphor, the phosphor emits light, causing a glow on the outside front of the tube.

The luminance signal controls the intensity of the electron beam. Because the amplitude of the luminance signal is always changing, the intensity of the beam is always changing. The horizontal and vertical sync signals operate magnets that are located around the neck of the tube. The vertical deflection magnets are controlled by the vertical sync signal, and the horizontal deflection magnets are controlled by the horizontal sync signal.

Of Lines, Fields, and Frames

One horizontal *line* of picture information is defined as the electron beam tracing the TV screen from left to right one time. At the end of the line, the TV set receives

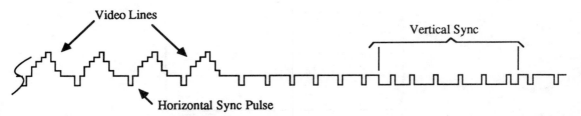

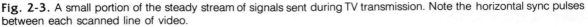

Fig. 2-3. A small portion of the steady stream of signals sent during TV transmission. Note the horizontal sync pulses between each scanned line of video.

a horizontal sync pulse, so the beam shuttles back to the left side, except circuits inside the TV set start the beam a little lower in the screen so that it doesn't trace the same spot over and over again.

The process repeats 262½ times before the electron beam reaches the bottom center of the screen. All of the lines traced so far comprise one *field*. Once the field has been traced, the vertical sync pulse pulls the electron beam back up to the top of the screen, (now starting at the center). The beam now sweeps over the screen another 262½ times, but this time, the new traces are between those previously made (see Fig. 2-4). When the beam reaches the bottom right corner of the tube, the vertical sync pulses again pull it back to the top and the process starts over again.

The two fields together make a *frame*. There are 60 fields per second and 30 frames per second. The pair of 262½-line fields comprise a total number of 525 lines per frame. The lines of the two fields are melded together with what's known as 2:1 interlace. The first field of the frame traces the odd numbered lines (1, 3, 5, etc.), and the second field of the frame traces the even numbered lines (2, 4, 6, etc.). The 2:1 interlace allows for less complex circuitry, yet the "refresh" rate of the screen (the time between each new successive frame) is high enough so that flicker is not visible.

Adding Color

In a black and white set there is one electron gun. In color sets, there are three electron guns, or at least one gun that produces three separate electron beams. The three beams are deflected together by the horizontal and vertical sync magnets and strike the face of the tube in a triangular pattern. Each beam hits a differently colored phosphor, either red, green, or blue. The intensity of the three colors at any one time determines the final color as seen on the outside of the tube.

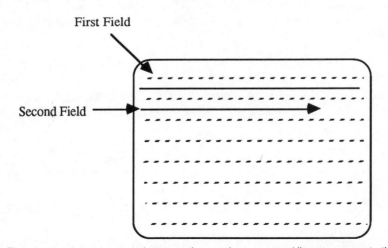

Fig. 2-4. The electron beam sweeps between the previous scanned lines to generate the second field of the video frame.

For example, if the green phosphors are lit but the others are not, you see an all-green picture. But if the red and blue phosphors are lit, but the green ones are not, you see purple. All the colors in the rainbow, and several million others, are derived from the three primary television colors.

DIFFERENCES BETWEEN BASEBAND AND RF SIGNALS

When TV picture information is combined with chroma and sync signals, it is called composite video. This is the lowest level of TV signal that you will encounter in home video setups. By contrast, professional broadcast systems use equipment where the luminance signal is often separate from the sync and chroma signals. (Exception: Super VHS VCRs have separate luminance and chrominance inputs and outputs. The Y-C output of the VCR is intended to mate with the Y-C terminals on a suitable television set. More about Super VHS and how it works is later in this chapter).

When composite video is not modulated on a carrier signal by any means, it is referred to as baseband video. The same is true of the audio portion of the program. Unmodulated audio is termed baseband audio. As you probably already know, baseband video and audio signals cannot be sent through the airwaves or down a cable. They must be carried by a modulated signal, one that combines both video and audio and is received over a specific TV channel.

Let's see how this works with the typical neighborhood channel 4 TV station. The signals are transmitted from the TV station antenna as radio frequency (rf) signals. Specifically, the channel 4 broadcast is AM and FM modulated in the 66-72 MHz range. The picture carrier is centered at 67.25 MHz, the color subcarrier is at 70.83 MHz, and the sound carrier is centered at 71.72 MHz.

To watch the station, you dial to channel 4 on your TV or VCR. The tuner in the TV or VCR then receives the signals sent at the frequencies listed above. After being processed by intermediate stages, the modulation carriers are stripped from the transmission and baseband audio and video result.

All tabletop VCRs have an rf output with an output frequency of either channel 3 or 4. You connect the rf output of the VCR to the VHF terminals of your TV, then tune the TV to channel 3 or 4.

If you have a VCR, you've probably noticed that it has both an rf output and separate video and audio outputs. The same sound and picture information is available at both outputs, it's just in different forms. The rf output is modulated to TV channels 3 or 4. The video and audio outputs carry baseband composite video and audio. You use one or the other output, depending on your system. Because of the extra processing stages inherent in rf reception and transmission, use baseband audio and video whenever possible.

BASIC VCR BLOCK DIAGRAM

The basic operation of all VCRs is the same. Figure 2-5 shows the block diagram of the major sections of a VCR, including record/playback heads, transport mecha-

HOW VCRs WORK

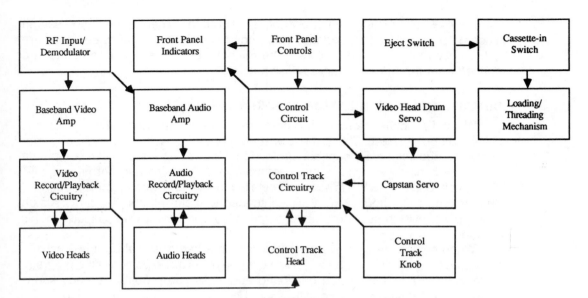

Fig. 2-5. A basic block diagram of the typical VCR. The central control unit in some decks is a microprocessor or one-chip computer.

nism, safety devices, and so forth. Note that not all VCRs have every block. For example, the control circuit is found only on models built after 1983 or 1984. Camcorders lack a timer and tuner.

CASSETTE LOADING

Home VCRs use a tape enclosed in a protective cartridge, also known as the shell or cassette. There are two types of cassette loading schemes: front and top.

Top Loading

Top-loading VCRs, where the tape is inserted into a lift mechanism on top of the deck, are no longer made (with one exception—the camcorder). To load the tape into the VCR, you press the EJECT button, which pops up the lift mechanism. You then slide the cassette in so that the reel hubs are on the bottom of the cartridge and the tape cover is in the front. After the tape is inserted, you manually press down on the cassette lift mechanism. Loading is complete. When loaded, the tape reels of the cassette rest over the supply and take-up spindles inside the VCR.

To unload the cassette, you stop the VCR (if it isn't already), and press the EJECT button. The cassette lift mechanism pops back up and you remove the cassette.

In top-loading Beta VCRs, the tape is automatically threaded around the tape transport mechanism and is unthreaded when you press the EJECT button. Because the tape is always threaded in a Beta deck, tape stress can cause buckling, which can lead to jams. That's why you should never leave a tape in a Beta deck when the machine is not actually in use.

Front-Load

Front-load VCRs are the mainstay, even though the loading mechanism is more complex. Front-loading allows you to fit the VCR in a tight slot in the front of the video system and not worry about obstructing the lift mechanism. With a top-loading VCR, you must leave the top free to access the cassette lift mechanism.

To load a tape, you insert it (hubs down, door to the front) into the loading slot (also called dock or opening). When the cassette has been inserted about mid-way, it triggers a leaf switch. This tells the VCR that a tape is inserted. The VCR then latches onto the tape and pulls it all the way into the cassette lift mechanism. After the cartridge is in the VCR, the cassette lift mechanism drops down. When loaded, the reels of the cassette rest over the supply and take-up spindles inside the VCR. The tape can now be threaded and played.

To unload the tape, press the EJECT button. The cassette lift mechanism lifts the cassette up and pushes it out the slot, where you can now retrieve it. On most VCRs, hitting EJECT will remove the tape even if the deck is playing. A mishap won't occur, however, because the VCR first stops the tape and unthreads it.

As with the top-load units, front-load Beta decks automatically thread the tape immediately after loading. The tape remains threaded until you eject the cassette.

TAPE THREADING

The transport mechanism of a VCR is considerably more complex than the transport of a reel-to-reel or cassette audio tape deck. In the old-model video tape recorders (VTRs), you were required to manually thread the tape around the heads and through the capstan and pinch roller. The cassette design of Beta, VHS, and 8mm decks relieve you from this requirement, so your hands never have to touch the tape.

VHS and Beta decks use dramatically different threading mechanisms, yet the end result is nearly identical. Let's take a closer look at the threading characteristics of each format.

Beta Threading

Figure 2-6 shows a simplified diagram of the threading mechanism of a modern Beta deck. When the cassette is loaded, the door on the cartridge that protects the tape is opened. That allows the VCR to gain access to the tape and pull it out of the cartridge.

When the tape is loaded into position (either with the manual cassette lift mechanism in top-loading machines or the automatic method of front-loaders), several threading guides and posts protrude inside the cassette area. These include a main

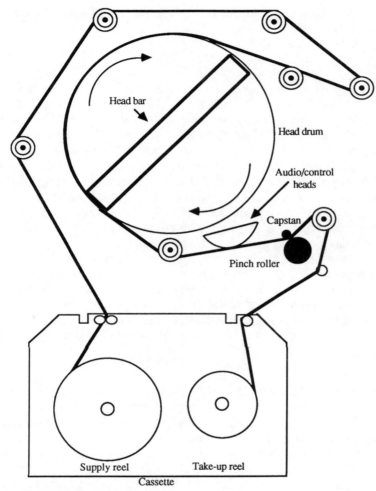

Fig. 2-6. A simplified view of Beta threading.

threading post and a capstan. The main (or lead) threading post is one of several posts mounted on a threading ring.

To initiate threading, the VCR activates a threading motor. This turns the threading ring, thus pulling tape out of the cartridge. The ring revolves around the video head drum and wraps the tape more than half way around it. Additional tape guide posts on the ring support the tape as it is threaded into position. When the ring finally stops turning, the tape is positioned against the erase, video, audio, and control track heads.

Some decks use additional rollers and guide pins. With a little imagination, you can visualize the threading layout as the letter "U." The "U-load" technique was also used in the ¾-inch video cassette system pioneered by Sony in the early 1970's (that system is still used professionally).

Note that early-model Beta decks used a different threading path layout. The major difference, as shown in Fig. 2-7, is that the tape is wound around the video head in the opposite direction. What is not as apparent is that the older threading system induced excessive tape stress and was prone to the problems of mis-threading. The newer Beta threading system poses little stress on the tape and the difficulties of mis-threading are greatly reduced.

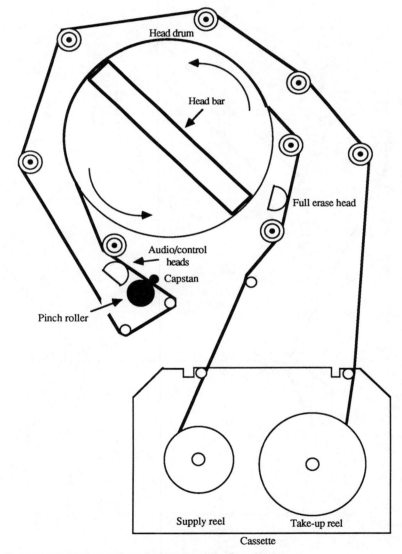

Fig. 2-7. How the tape was threaded in the older Beta decks.

VHS Threading

Since their introduction, all VHS decks have had the same threading procedure. When the tape is loaded into position, two tape-threading guides, a tape tension lever, and a capstan protrude into the cassette area, as shown in Fig. 2-8. Threading is accomplished by the two tape-threading guides. These guides, which are mounted on either side of the video head drum, pull the tape out of the cartridge and wrap

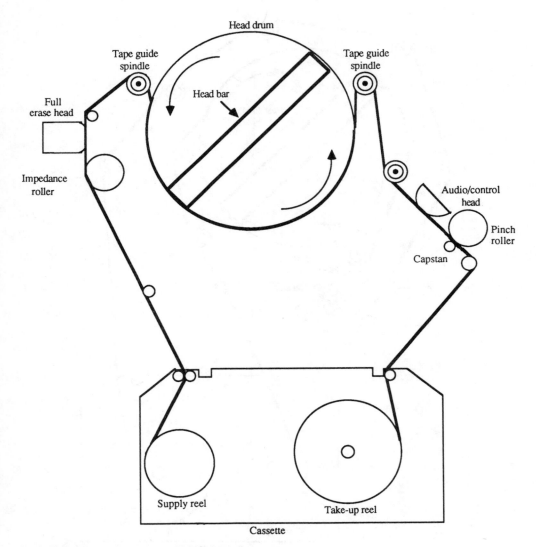

Fig. 2-8. A simplified view of VHS threading.

it around the drum. The erase, audio, and control track heads are positioned so that the tape is pressed firmly against them once threading is complete.

During threading, the pinch roller comes into contact with the tape and capstan and acts as a pressure roller for tape transport (see the following section). The tape tension lever, mounted on the left side of the transport area, swings to the left. The lever helps maintain even back-pressure on the tape, and prevents the tape from spilling out into the transport area.

The VHS threading path layout takes on an "M" shape, so VHS decks are said to use "M-loading." VHS-C and 8mm decks use similar loading techniques.

Unthreading

The threading mechanism in the VCR works in reverse to unthread the tape. In Beta decks, the threading ring unwinds while the supply spindle back-spins to take up the excess. In VHS decks, the two threading posts retract and the supply spindle back-spins to remove slack.

TAPE TRANSPORT

VHS and 8mm decks thread the tape only for play and record (plus fast-scan or search modes). The tape is unthreaded for fast-forward and rewind. Conversely in Beta VCRs, the tape is always threaded until the EJECT button is pressed.

With the tape threaded, the deck is ready for recording or playback. Several things occur within the tape transport mechanism during recording and playback:

- ✦ The capstan and pinch roller pull the tape through the transport and past the heads.
- ✦ The take-up spindle winds the tape that has passed through the transport onto the take-up reel.
- ✦ The video head drum spins at 1,800 revolutions per minute (30 revolutions per second) to record and play back video.

Head Drum

The video head drum spins under propulsion from a direct-drive motor (in some early VCRs, the drum was operated by an AC or DC servo-controlled motor). This direct-drive motor contains no brushes, just coils and magnets. The coils in the motor (usually three) are switched on and off via a servo circuit.

Precise timing of the switching is accomplished by a novel feed-back mechanism provided by a set of Hall-effect integrated circuits (ICs). The Hall-effect ICs, which are the electronic equivalent of reed switches, are magnetically sensitive and can detect the presence or absence of a magnetic field. In older Beta and VHS models, coils and magnets are used instead of Hall-effect sensors.

The Hall-effect ICs generate two types of pulses, *PG* and *FG*. PG stands for pulse generator; FG stands for frequency generator. The PG signal, typically at 30 Hz, identifies the position of the drum as it spins. The rate of the FG signal, most often

HOW VCRs WORK

180 Hz, indicates the speed of the video head drum. As the drum rotates, the Hall-effect sensors relay the position and speed of the drum to the servo.

The servo, which is clocked to a precise standard, increases or decreases the driving signal to the video head drum motor. The driving signal is pulse-width modulated—often referred to as PWM. In the PWM system, the same voltage is always applied to the motor, but the duration of the pulses are shorter or longer, depending on whether the rotation of the drum is to remain the same, speed up, or slow down. Figure 2-9 shows how PWM works.

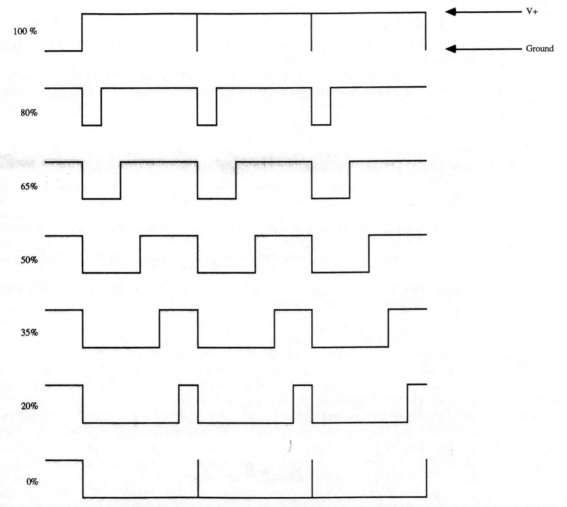

Fig. 2-9. Pulse width modulation retains the same voltage levels but varies the amount of time the levels remain in one state. The longer the voltages are on, the greater the duty cycle.

During recording, the servo is governed by a 60 Hz pilot signal. On the old VCRs, that signal was usually derived from the ac line frequency; now the signal is invariably created by the 3.58 MHz color subcarrier quartz crystal (divided down to 60 Hz by a counter or set of flip-flops). With each revolution of the video head drum, a short tone or pulse is placed on the control track by the control track head. The control track head records 30 pulses per second. During playback, the servo is maintained by the signal encoded on the control track as well as a clock circuit on the VCR.

The Role of the Control Track Head

The control track signal serves three basic functions:

- The signal identifies the speed of the recording so that the VCR can adjust its playback speed accordingly (recording speed is set by a switch but playback speed is automatic).
- The signal maintains proper playback speed even when the tape stretches or shrinks.
- The signal ensures that the video heads trace the video tracks made during recording.

Without the control track, the VCR would not be able to operate. If the control track becomes damaged or the control track head becomes dirty, proper recording and playback is impossible.

Beta and VHS VCRs are equipped with a tracking control knob to fine-tune the timing between the control-track pulses and the scanning of the video heads. In normal operation, the control is centered in its detent position. Tracking errors, which appear as lines in the top and/or bottom of the picture, are corrected by adjusting the tracking control to retard or advance the timing between the control-track pulses and the motion of the video heads. The 8mm VCRs lack a control-track knob; pilot signals are recorded on the tape that serve to identify the tracks made by each head during recording.

Capstan

The capstan is driven via a belt or roller by a precision DC motor, as depicted in Fig. 2-10. The speed of the capstan determines the rate at which the tape is pulled through the transport. If the speed varies, both the audio and video are affected. *Wow* and *flutter* are induced in the audio when the capstan speed is not properly regulated. The picture might roll or jump, and heavy lines might appear all through the screen. If the speed variations are great, the picture can be muted completely by the VCR circuits, and you'll see nothing.

Small speed imperfections are smoothed out by the capstan flywheel, a heavy metal wheel that uses inertia to keep the capstan speed more-or-less constant. Precise control of the capstan is maintained via electronic means.

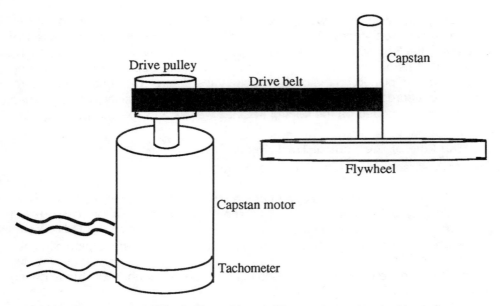

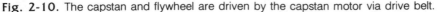

Fig. 2-10. The capstan and flywheel are driven by the capstan motor via drive belt.

The speed of the capstan is controlled electronically by a two-element process: frequency-related FG tachometer pulses, generated by the capstan, and position-related PG timing pulses, originally generated by the video head drum. The PG pulses from the video head drum serve as a reference, and the FG pulses generated by a tachometer built onto the capstan serve as feed-back.

At any given recording or playback speed, the VCR examines the rate of PG pulses coming from the video head drum (which are always the same) and compares them to the rate of FG pulses coming from the capstan (which change depending on the tape speed).

As with the video head drum, the VCR changes the speed of the capstan motor not by varying the voltage but by lengthening or shortening the time (or duty cycle) the capstan motor receives full voltage.

Note that testing the capstan motor input voltage with a meter will yield different voltages, depending on the playback speed. Why? The volt-ohm meter cannot react to the fast PWM pulses delivered to the motor, so it averages them out. During EP play, when the tape is traveling the slowest, the meter reading might be something like .5 volts. The actual voltage to the motor could be on the order of 12 volts, but the average voltage is far below because the motor is receiving power for a short period of time.

During SP play, however, the motor receives power for a greater period of time, so the average voltage can double or triple. In fast-scan, the voltage can swing close to the full 12 volts.

Keep this in mind when testing the capstan in your VCR. You may test overall operation by connecting the leads of the meter to the power terminals, but you have no idea if the motor is receiving the correct duty cycle for that given speed.

Take-up Spindle

During record and playback, the take-up spindle turns to wind the tape onto the take-up reel inside the cartridge. In some VCRs (most notably those made by Mitsubishi), the take-up spindle is driven by a separate motor. The speed of the motor is regulated to keep the tape taut. If the motor stops, or doesn't turn fast enough, the capstan will pull too much tape through the transport and excess tape will accumulate in the VCR, causing a jam.

In most VCRs, the take-up spindle is driven by a roller. In turn, the roller, as shown in Fig. 2-11, is attached to a belt that connects to the capstan motor. This not only shaves a few dollars off the manufacturing price of the VCR, but it helps ensure that the take-up reel travels in proportion to the capstan. The faster the capstan turns, the faster the take-up spindle turns.

As with a reel-to-reel audio deck, the free-running speed of the take-up spindle is greater than the speed of the capstan. If you release the pressure on the capstan exerted by the capstan pinch roller, the take-up spindle will pull the tape through the transport at increased speed. In fact, this is how most decks accomplish fast-forwarding. Because the capstan is not limiting the maximum speed of the tape, the tape is wound at the free-running speed of the take-up spindle.

Note that the torque of the take-up spindle during normal recording and playback is controlled by a torque limiter. This limiter, usually a pad or brake, prevents the

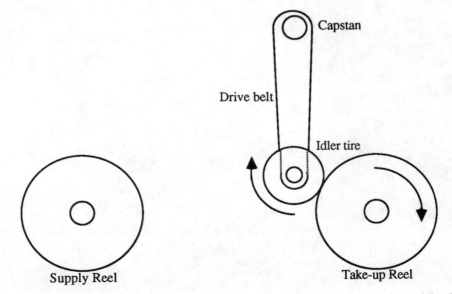

Fig. 2-11. In most VHS VCRs, the take-up and supply reel tables are driven by an idler tire or roller, which is directly linked by a belt to the capstan motor.

take-up spindle from exerting too much force on the tape, and therefore stretching or breaking it.

Supply Reel Spindle

All VCRs have a supply-reel spindle as well. The spindle is normally unpowered during recording, playback, and fast-forward but is powered during reverse play (certain VCRs only), reverse search/fast-scan (or "review"), and during rewind. The same belt and roller that drive the take-up spindle also drive the supply spindle. An idler tire swings back and forth, as shown in Fig. 2-12, to deliver power to either one spindle or the other.

It should be noted that the condition of this idler tire is critical to the performance of your VCR. Almost all VCRs use this tire (even the newest state-of-the-art models). If the tire becomes worn, glazed, or dirty, proper speed of the supply and take-up reels is not maintained. That invariably leads to tape spillage.

Fig. 2-12. Close-up view of the idler tire (attached to a pivoting arm) in a VHS deck.

RECORDING AND PLAYBACK OPERATIONS

VCRs use magnetic heads to record and play back video, audio, and control information. Let's take a few moments to review how magnetic heads operate.

The Role of Magnetic Heads

The basic idea behind magnetic head operation is simple: To record, an ac signal is applied to a coil that is wound around a metal core. The core is split down the middle to create an open-air gap, like that in Fig. 2-13. Without the gap, the head would not operate.

A magnetic *flux* (field) is induced at the gap, and changes as the input signal to the head changes. The intensity and magnetic polarity of the flux varies, depending on the intensity and polarity of the incoming ac signal. To make a permanent copy of the signal, a tape coated with a magnetically-sensitive ferrous substance (usually ferric oxide) passes under the head. The magnetic flux produced by the head is imparted to the tape.

For playback, the tape is passed under the head, but this time, no signal is applied to the coil (that would re-record another signal). The magnetism of the tape causes minute electrical changes in the coil of the head. These changes are amplified and sent to the appropriate circuitry for playback.

Though there is more to magnetic recording and playback to this (such as the addition of a high frequency "bias signal" to faithfully record baseband-level audio and control-track signals), the discussion is sufficient for our purposes. The idea

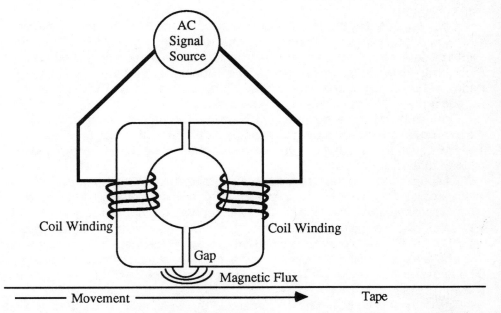

Fig. 2-13. Magnetic heads record a signal by passing an alternating current through a coil winding.

is that once the tape is magnetized by the heads, the magnetism remains (more or less) until something else is recorded onto the tape, obliterating the previous recording.

Signal Conversion for Recording

The TV spectrum covers a considerable frequency range of 4.2 MHz (that's just for the picture; the range is 4.5 MHz with the audio signal added). Consumer-level magnetic tape does not possess the frequency response to accurately record information over such a wide range, so to place audio and video onto a VHS, Beta, or 8mm cassette, the signals must be modified. The modification is extensive and varies slightly depending on the format. We'll address the VHS format here, then discuss the differences found in Beta and 8mm VCRs.

Luminance and Chrominance Conversion For Recording

After the rf signal is received by the VCR's tuner, it is demodulated (stripped of the modulation carriers), the video portion is sent to the video circuits, and the audio portion is sent to the audio circuits. With few exceptions (as with Beta hi-fi decks, detailed subsequently), the audio and video never meet again until the taped program is played back and viewed on your television set.

The composite video is then passed through two filters, as shown in Fig. 2-14. One is a low-pass filter, the other a band-pass filter. The low-pass filter passes just the luminance component of the video signal and rejects the chrominance component, which is situated at approximately 3.58 MHz. Let's follow the luminance signal for now.

Once separated by the filter, the luminance component (or channel) is amplified. The signal is equalized depending on the recording speed (SP, LP, or EP). Equalization is important to prepare the signal for optimum recording at the desired record/playback speed. Pre-emphasis and clipping is added to increase high-frequency response and signal-to-noise ratio, and to prevent the signal from exceeding defined upper and lower limits.

The composite video signal is then FM modulated with a carrier signal so that the total deviation of the video signal spans a 1 MHz range between 3.4 to 4.4 MHz. The FM signal, along with the lower luminance channel sideband, is amplified and applied to the video heads.

Of course, this isn't the whole story. After going through the low-pass filter and video amplifier, the horizontal synchronization signals are detected and separated by a sync separator. One side of the sync separator is applied to a color sync generator, discussed later, and the other to a carrier generator and phase rotator.

The carrier generator creates a 629 kHz cw signal that feeds the phase rotator. The phase rotator is used to rotate the vector or phase of the color subcarrier signal 90 degrees for each video line. The rotation helps prevent crosstalk between adjacent lines of video.

The chrominance component is separated by the band-pass filter and applied to a color amplifier. The output of the amplifier is controlled by an automatic color

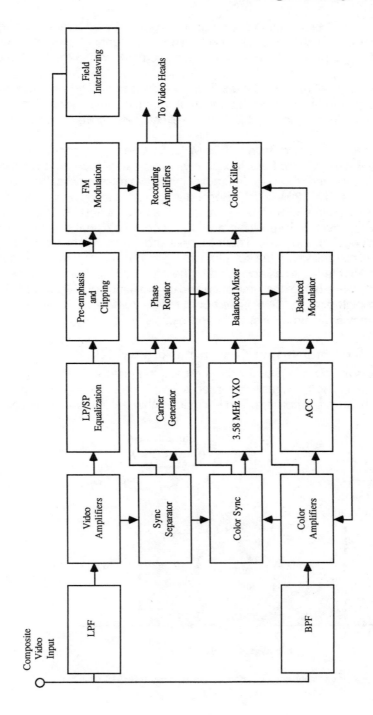

Fig. 2-14. Block diagram of the VHS recording circuit.

corrector, or ACC. The color section uses a gating signal created by the horizontal sync rate to separate the chrominance signal. When the signal is present, it turns the color killer off. That passes the color burst signal created within the VCR to the heads.

When the signal is not present, the color killer circuit is turned on. That prevents the color burst signal from reaching the heads. If the color burst signal is present during a monochrome recording, an objectionable effect called *moire* or "color-crawl" can occur.

The color sync circuits also drive a 3.58 MHz variable crystal oscillator (VXO). The oscillator generates a color-burst-locked cw signal that is fed into a balanced mixer. The other input of the mixer is the output of the phase rotator, described earlier. The output of the mixer is a 4.2 MHz color signal. It is applied to a balanced modulator, which is also driven by the 3.58 MHz signal from the color amplifier. The product is a difference heterodyne that results in a 629 kHz chroma signal sideband. This 629 kHz signal is coupled with the luminance signal and applied to the video heads.

The technique of transposing the color to a frequency below the luminance signal (or channel) is often referred to as color-under (the process is technically known as *heterodyning*). The waveform of the luminance and chrominance signals, as recorded by the VCR, is depicted in Fig. 2-15. Note how it differs from Fig. 2-2, earlier in the chapter, which shows the broadcast television picture spectrum.

Luminance and Chrominance Conversion For Playback

The process is almost reversed for playing back the picture. Refer to Fig. 2-16 as you read the text. The previously recorded signal is detected on the tape and amplified by the head pre-amplifiers. The luminance channel is then processed by

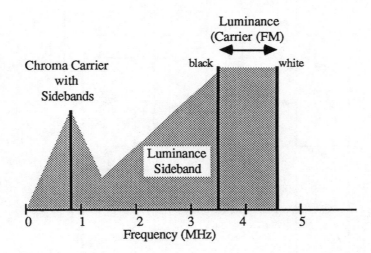

Fig. 2-15. The video signal frequencies recorded on tape by the VHS system.

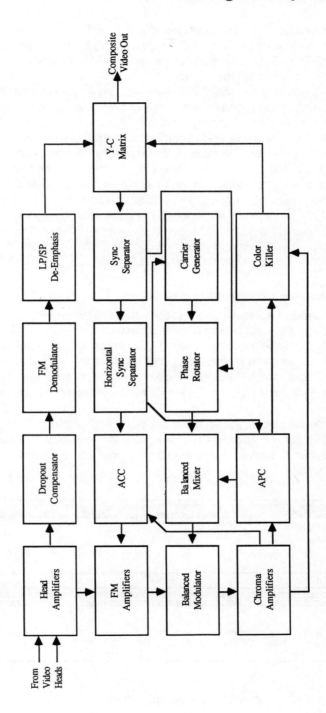

Fig. 2-16. Block diagram of the VHS playback circuit.

a drop-out compensator that duplicates the information from previous lines to fill in blank areas of tape (caused by a loss of oxide coating on the tape). The drop-out compensator (DOC), which is standard equipment on all VCRs, limits the appearance of black and white specks on the screen.

Once compensated, the signal is FM demodulated and is returned to its original baseband component. De-emphasis removes slight signal alterations that were made depending on the speed of recording. The output is then applied to a Y-C matrix, which combines the luminance and chrominance channels together.

The 629 kHz color-carrier rotating sidebands are sent to an rf amplifier and balanced modulator. The rotation is removed thanks to sync separator circuits driven by the luminance component. The color signal is then amplified by a chroma amplifier that drives the color killer, automatic phase control (APC), and automatic color control (ACC) sections. The output of the color killer section when the color burst signal is present is a corrected (non-rotating) color carrier. The output of the Y-C matrix is baseband video and is routed to the VIDEO OUT terminals of the VCR or sent to the rf modulator section of the VCR.

Audio Signal Recording and Playback

After the baseband audio and composite video signal are separated, the audio portion is sent to the audio recording circuits. These circuits work identically to those found in a reel-to-reel tape recorder. The signal is mixed with a high-frequency bias current (which improves linear response) and is recorded on tape as a baseband signal. The linear audio track is along the top of the tape. In the case of linear stereo (in VHS decks and European Beta decks) there are two discrete tracks, as shown in Fig. 2-17.

For playback, the heads pick up the audio track and amplify it. If noise reduction is used, such as Dolby, the audio is processed through the noise reduction circuitry. The amplified audio is applied to the AUDIO OUT terminals of the VCR and also is sent to the rf section.

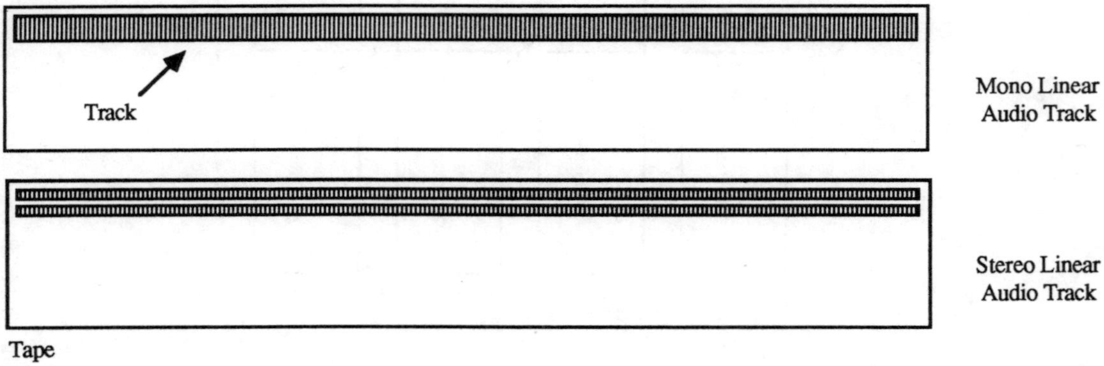

Track

Mono Linear
Audio Track

Stereo Linear
Audio Track

Tape

Fig. 2-17. Monophonic and stereophonic linear sound tracks.

Control Track Signal Recording and Playback

The vertical sync signal is not recorded with the rest of the video because VCRs don't require vertical sync the same way as TV sets. The equivalent of vertical sync is recorded on the control track heads. As discussed earlier, the 30 Hz signal is used to properly time the video heads in relation to the tracks on the tape.

Beta VCRs

The process of recording and playing back video, audio, and control track information is similar with Beta models. Figure 2-18 shows a block diagram of the Beta recording function while Fig. 2-19 shows a block diagram of the Beta playback function.

The biggest differences in the recording process are the FM deviation frequency for the luminance channel, the chrominance subcarrier frequency, and an alternate-line phase inverter circuit. In Beta decks, the luminance channel is frequency modulated to between 3.5 and 4.8 MHz—a 1.3 MHz range (this wider bandwidth is what has always given Beta the edge over VHS). The color channel is down-converted to 688 kHz. On each line of video, the phase of the color burst signal is inverted 180 degrees. As with the VHS system, the phase inversion reduces crosstalk between adjacent video lines.

The system is nearly reversed for playback. As with VHS, the Beta luminance signal is processed through a drop-out compensator. The alternate-line color subcarrier signal is corrected, up-converted to 3.58 MHz, and mixed with the luminance component in the Y-C matrix circuit.

8mm VCRs

VCRs in the 8mm camp are, in many ways, scaled-down versions of VHS machines. The same color-under recording technique used in VHS and Beta is employed in 8mm except that the chrominance carrier is centered at 719 kHz. An automatic track finding (ATF) signal, recorded 14 dB below the level of the chrominance carrier, eliminates the need for manual tracking control adjustment. The ATF system uses four pilot tones to mark successive tracks. During playback, the ATF circuitry identifies the tones and continually adjusts the tracking so that the tones are steady and equal.

VIDEO HEAD GEOMETRIES

VCRs use a minimum of two video heads to record and play back picture information. Each head records a single TV field in one pass, so two adjacent tracks on the tape comprise an entire video frame. In both VHS and Beta systems, the heads are mounted 180 degrees apart, and only one head touches the tape at any time (in VHS-C camcorders, the tape wraps around the head 270 degrees and two heads record on the tape at the same time).

Head Gap Size

The size of the gap found in all magnetic recording heads varies depending

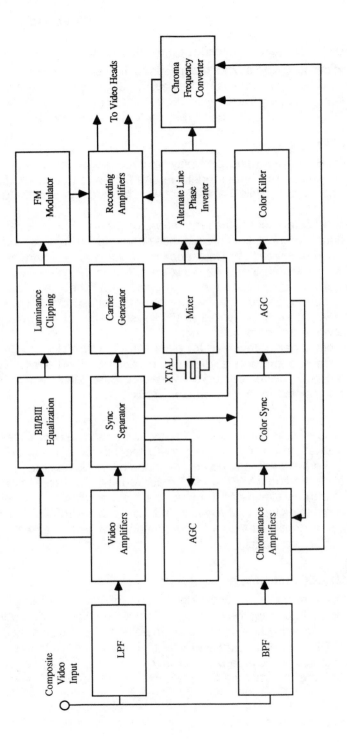

Fig. 2-18. Block diagram of the Beta recording circuit.

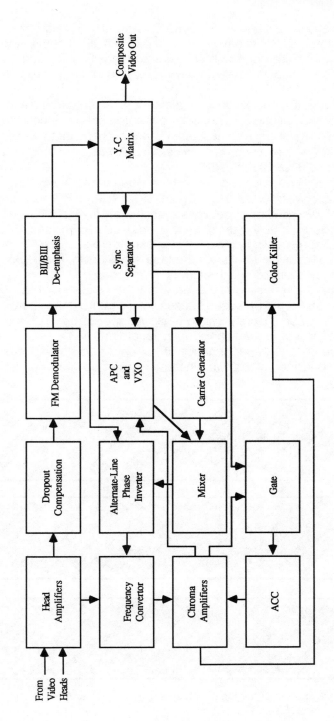

Fig. 2-19. Block diagram of the Beta playback circuit.

on the application and varies from about 28 microns to 90 microns wide. The video heads in typical VHS decks have a head gap of only 38 microns—too small for you to see without a microscope or high-powered magnifier. The gap in conventional Beta video heads is even smaller, 30 microns, and in 8mm decks, the gap is 20.5 microns.

VCRs use head gaps of various sizes to increase performance at slower record and playback speeds and to accomplish noise-free special effects. Wider heads record a stronger signal, and are desired over narrow head gaps. But as you will soon see, a problem occurs when using wide heads to record and play back video at the slower LP, EP, or BIII speeds.

Figure 2-20 shows the geometry of the video tracks layed down by the video heads with a conventional VHS VCR at SP, LP, and EP speeds. VHS design parameters allows for 58 microns between each video track at SP speed. The spacing decreases proportionately at the slower speeds—29 microns and 19 microns for LP and EP speeds, respectively. At the fast speed, there are guard bands as wide as 28 microns between the tracks. As the tape slows down, the tracks get closer together and they overlap by as much as 11 microns.

Obviously, if the deck is equipped with wide heads, playback and recording will be satisfactory at the SP speed because of the guard bands between the tracks. But operation will be severely limited at slower speeds because the tracks will almost stack on top of one another during recording.

To provide optimum performance at both fast and slow speeds, many VCRs use four heads instead of just two. The gap is wider on one side of the heads. The

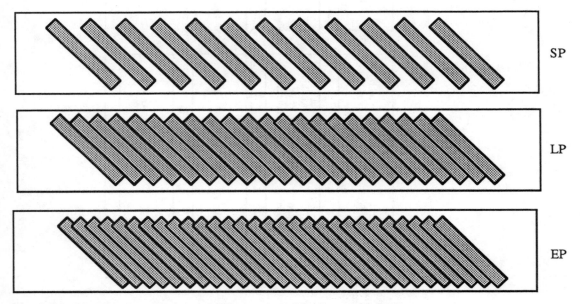

Fig. 2-20. Track spacing at the three VHS speeds of SP, LP, and EP (SLP).

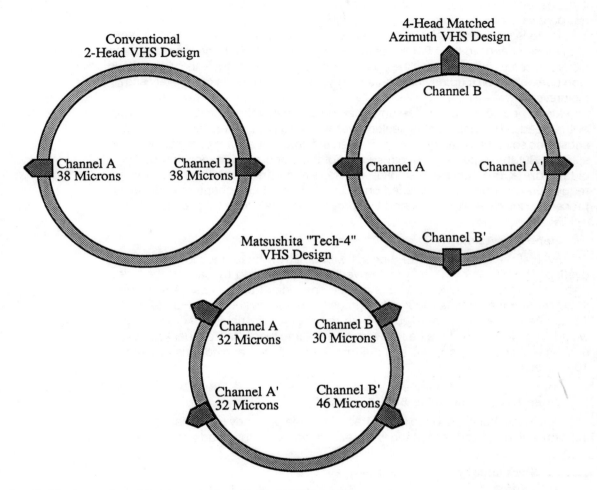

Fig. 2-21. A sampling of head geometries of typical VHS machines.

wide set is designed for SP speed; the narrow set for LP and EP speed. The actual size of the head gap varies, depending on the make and model. Figure 2-21 shows the head geometries for popular two- and four-head VHS designs from JVC and Matsushita. Note than the head gap is not always the same for opposite heads in the same set. This provides the deck with better performance during playback and with certain special video effects, such as still-frame and slow motion.

When Heads Become Dirty

When audio heads acquire an excessive build-up of tape oxide and other contaminants, the sound output is reduced and might be fuzzy and indistinct. But if you haven't trained yourself to listen for the effects of dirty audio heads, you might not be aware of it.

There is no such blissful ignorance with dirty video heads. When one or more of the video heads in the VCR become dirty or "clogged" with foreign matter, the picture becomes washed out. If the accumulation of junk is excessive, the video heads won't be able to pick up anything on the tape, and you will see a completely snowy picture.

Many people mistakenly assume that foreign material gets caught in the gap of the heads. The gap itself is sealed over with epoxy or other hard, non-ferrous substance so nothing can get inside. Rather, the contamination is completely on the surface. But the extremely smooth surface of the video head means that particles of tape oxide, dust, cigarette smoke, and so forth, can adhere like glue. All VCRs require that their heads be cleaned on a regular basis. Later chapters address the procedure for cleaning heads and the proper care that must be exercised.

Opposing Azimuth

Even at the slow LP and EP speeds with a narrow head, the video tracks still overlap. The effects of this inevitable overlap are precluded by canting the video heads at opposing angles. In VHS systems, one video head is tilted +6 degrees to center and the other is tilted −6 degrees to center, as shown in Fig. 2-22. The 12 degree difference in so-called azimuth reduces or eliminates the crosstalk that would otherwise occur if the heads were on the same plane. In Beta decks, the azimuth is +7 and −7 degrees, and in 8mm the heads are angled plus and minus 10 degrees.

Video Heads and Noise Bars

During playback, the video heads trace the same path they took when recording the signal. If the speed of the tape is increased, however, the heads no longer can

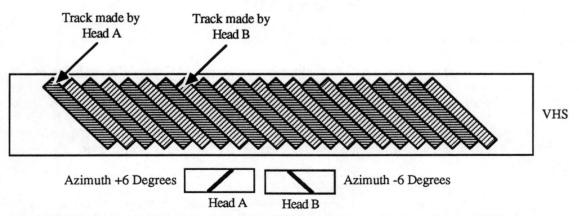

Fig. 2-22. Alternating video tracks are recorded at different angles to minimize cross-channel interference.

trace over just one track at a time because the angle of the heads has changed in relation to the tracks recorded on the tape. The result: the heads cross into adjacent tracks. In one pass, the video head might scan a portion of two, three, or even more tracks. The effect you see on the screen is a picture composed of stripes from each successive video track. The stripes are separated by "noise bars," as shown in Fig. 2-23.

These bars can never be totally eliminated in consumer VCR equipment, but they are reduced with wide heads or when recording with regular heads at slower speeds. With extra wide heads at SP speed, for example, there is little *guard band* space between the tracks. But the guard bands are wider when recording at SP speed with narrow heads. It is the guard bands, which are unrecorded areas of the tape, that are the cause of the most objectional noise bars.

The noise bars can be heavier when a tape recorded with a deck that uses narrow heads is played back even on a machine that uses wide SP heads. The bars are thicker because the wider heads pick up extra guard band area.

Special Three- and Five-Head Systems

A number of VCRs have an extra third or fifth head that is used for noiseless

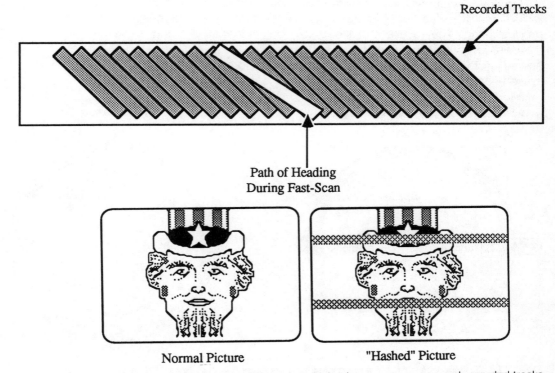

Recorded Tracks

Path of Heading During Fast-Scan

Normal Picture

"Hashed" Picture

Fig. 2-23. The path of the video head during fast-scan playback crosses over several recorded tracks, resulting in a banded or "hashed" picture.

freeze-frame or slow motion. The extra head, which is often referred to as the "trick" head, is not used for recording and normal playback, but only to deliver the cleanest possible picture in freeze-frame and slow motion operation. Figure 2-24 shows the geometry of three-and five-head systems and how the heads are used for the various operations.

This extra head can be a source of confusion during cleaning and maintenance. If the playback heads become dirty with tape oxide build-up or other foreign matter, you won't be able to see a picture. But should you place the deck in freeze-frame or slow motion mode, the picture might return. The reason for this is obvious: different heads are being used for standard and special effects playback. Always be sure to thoroughly clean all of the heads before assuming that something serious is wrong with the VCR.

Flying Erase Heads

In conventional VCRs, the full-erase head is used to totally "scramble" the magnetic coating on the tape so that a new signal can be properly recorded. Without full erasure, a portion of the old program might still remain. You might hear a faint background sound in the audio portion, and the video could appear washed out, grainy, or contain wavy, colored lines (called moire).

VCRs with video dubbing capability use the video heads to re-record over old programming. The results are acceptable but marginal. A flying erase head is mounted just before the video head in the head drum assembly (see Fig. 2-25) and erases just the video tracks and prepares them for proper re-recording. With flying erase heads, the video dub looks better and is not marred by moire patterns.

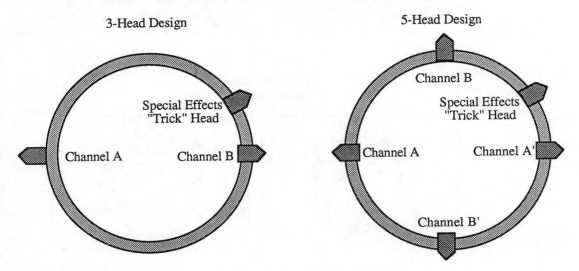

Fig. 2-24. Special three- and five-head designs used in some VHS decks.

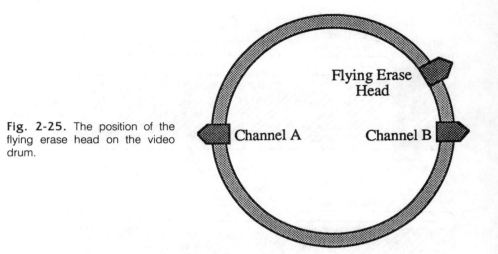

Fig. 2-25. The position of the flying erase head on the video drum.

Flying erase heads are often found in camcorders, which are used for start-and-stop shooting. The flying erase head provides for clean, glitch-free transitions between scenes, as shown in Fig. 2-26, even if you record both picture and sound over a previous segment. In most consumer VCRs, the flying erase head deletes two video tracks at the same time.

Head Switching

During recording, the signal is fed alternately to the heads by a head-switching circuit. The head is "turned on" as it touches the tape and makes its half-circle pass around the head drum. Just before the head is about to leave the tape, it is turned off and the signal is applied to the opposite head. The video waveform for the head switching is shown diagramatically in Fig. 2-27.

INSIDE HI-FI AUDIO

The first Beta and VHS VCRs recorded monophonic sound only. In 1982, Sony came out with a novel stereo system for their Beta decks that recorded high-fidelity stereophonic sound. Not to be outdone, JVC soon developed a stereo hi-fi system for their competing VHS decks.

Although both Beta and VHS hi-fi decks share similar audio specifications, the process of recording and playing back the audio is completely different between the two systems. Both VHS and Beta hi-fi offers a dynamic range (soft to loud) of about 80 dB. Conversely, the linear audio track on conventional VCRs delivers a dynamic range of no more than 50 dB.

Beta Hi-Fi

VCRs use rotating video heads to record large amounts of information in a relatively short space of tape. In a VCR, the tape literally crawls along, but the helical

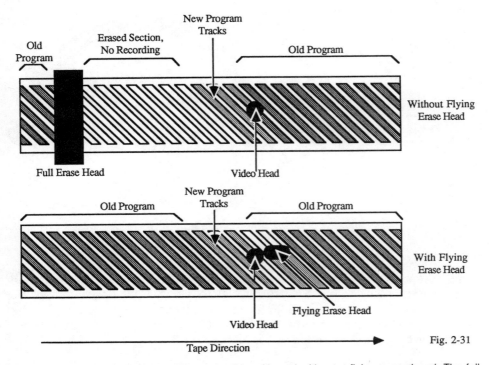

Without Flying
Erase Head

With Flying
Erase Head

Fig. 2-31

Fig. 2-26. The effects of insert editing recording with and without a flying erase head. The full-erase head causes a large gap of unrecorded material between old and new programs; the flying erase head leaves no such gap.

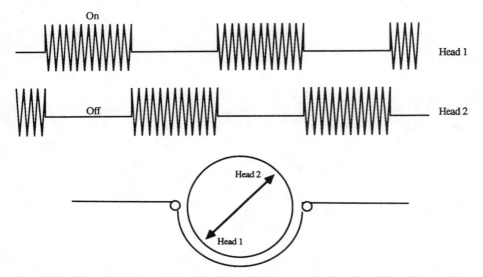

Fig. 2-27. The action of switching the video heads on and off as the video drum rotates.

scan system enables the video heads to trace a far greater surface area of the tape. Similarly, high-fidelity sound is not possible using the stationary audio heads in the VCR because the tape moves too slowly. VCR engineers reasoned that if they improved video performance by using rotating video heads, they could do the same thing by using rotating audio heads.

Using a second set of heads would be the expensive way to add high-fidelity to a VCR. Therefore, the hi-fi sound is included with the video and recorded with it on the tape. Figure 2-28 shows the standard Beta recording format, discussed earlier in this chapter, and the Beta hi-fi recording format. Notice two changes: the luminance carrier is moved up 400 kHz (in Super Beta decks, discussed

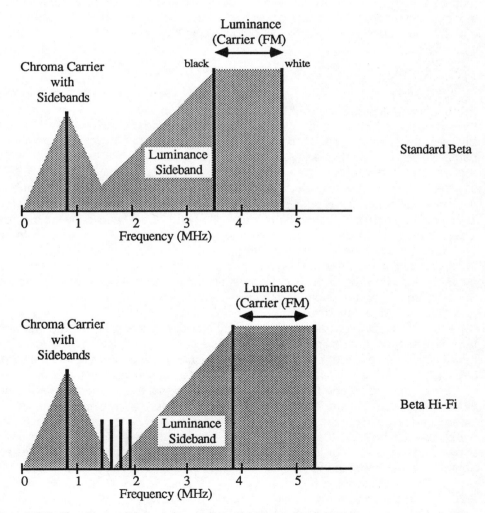

Fig. 2-28. Signal layout for beta standard and hi-fi recording formats.

subsequently, the luminance channel is moved up 800 kHz). That opens more space between the luminance sideband (the diminishing harmonics of the luminance carrier) and the chrominance carrier.

The Beta hi-fi audio signal is frequency modulated and placed in the nook between the luminance sideband and chroma carrier. The sound is modulated onto FM carriers so that for one video field, the frequencies are 1.38 MHz (left channel) and 1.68 MHz (right channel). During the next field, the FM carriers are placed at 1.53 and 1.83 MHz. The difference in frequencies between the fields helps reduce crosstalk.

Although the Beta hi-fi system moves the recorded signal around a bit, tapes recorded in hi-fi are still playable on most decks that lack the hi-fi capability. A monophonic rendition of the sound is placed on the audio track and the deck "seeks out" the luminance carrier and locks onto it, even though it is 400 Hz higher than it should be.

VHS Hi-Fi

JVC had considerable trouble developing a competing hi-fi audio standard for their VHS decks. The standard VHS recording format does not allow enough space between the luminance sideband and the chrominance carrier, even if the luminance carrier could be moved up a few notches or two. In fact, VHS decks cannot tolerate a change in frequencies.

This means that VHS VCRs cannot use the video heads to record hi-fi audio. A second set of heads, as shown in Fig. 2-29, are added to the video head drum. The heads operate in the same manner as video heads. One head in the set records a field's worth of audio information, the next head records another, and so forth. The gap of the hi-fi heads can vary from about 26 microns to 42 microns.

In the VHS hi-fi system, the audio signals are recorded first, then the video signals are recorded. Normally, over-recording with the video heads would lead to complete or partial erasure of the audio track, but in practice, this does not happen—thanks to an old recording technique.

It's been known for many years that low frequency signals penetrate into the tape farther than high frequency ones. Since the hi-fi audio portion is lower in frequency than the 3.4 to 4.4 MHz luminance carrier, it travels deeper into the tape. JVC calls this technique "depth mutliplex recording." It sounds complex but it isn't.

During recording, the hi-fi heads record their signal a split second before the video heads record theirs. The audio signal goes deeper into the tape, so when the video heads pass by, most of the already-recorded sound information remains. To avoid crosstalk between the audio and video carriers, the hi-fi audio heads are tilted at plus and minus 30 degrees. The extreme azimuth also reduces crosstalk between right and left stereo channels.

The hi-fi heads record the audio on FM carriers, as shown graphically in Fig. 2-30. The frequencies are 1.3 MHz for the left audio channel and 1.7 MHz for the right audio channel. Compatibility with non-hi-fi decks is retained because the sound

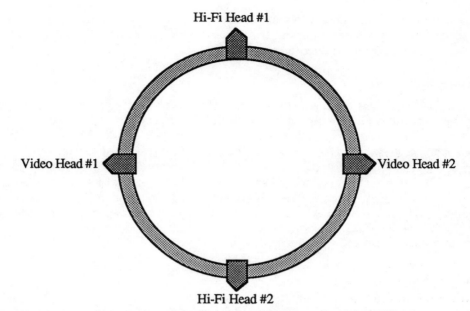

Fig. 2-29. The position of the video and audio hi-fi heads on the typical VCR.

is also recorded on the linear audio tracks. Many stereo hi-fi VCRs also have linear stereo audio heads.

Hi-Fi Limitations

Because the hi-fi audio is recorded with the video track, effects such as audio dubbing are not possible. There is no way to selectively erase the audio portion of

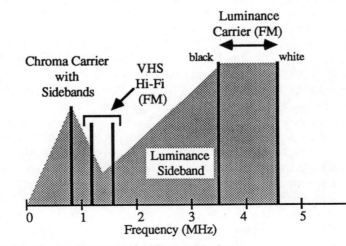

Fig. 2-30. Signal layout for the VHS hi-fi recording format.

the program and replace it with another track. However, many VHS hi-fi machines also record stereo audio onto the two linear audio tracks. These tracks can be selectively erased and re-recorded.

SUPER BETA

Super Beta adds another 20 percent detail to the video picture. It does this mainly by relocating the luminance carrier (some additional circuitry is also employed). Instead of positioning the luminance carrier between 3.5 and 4.8 Mhz (a 1.3 MHz deviation), as in standard Beta units, Super Beta decks translate the luminance carrier up 800 kHz to 4.4 through 5.6 MHz (a 1.2 MHz deviation). Resolution is increased from 250 horizontal lines (standard Beta) to about 290 horizontal lines (Super Beta).

Compatibility with most older Beta decks that lack Super Beta circuitry is maintained because Beta decks can seek out the higher luminance carrier frequency and lock onto it. However, some early-model Beta decks have been found that do not successfully track the higher frequency.

SUPER VHS

VHS VCRs cannot cope with deviations in frequency standards, so to apply the same technique used in Super Beta with VHS units would impair compatibility. Decks that lack the "Super VHS" circuitry would not be able to play back tapes recorded on a Super VHS deck.

JVC finally took the plunge in 1987 and revealed a Super VHS deck. It employs the same general technique found in Super Beta decks, plus it adds a few other capabilities as well. From the outside, a Super VHS deck looks the same as a conventional VHS video recorder, and has similar features, such as three-speed recording (most models), wireless remote controls, fast scan in reverse and forward, and freeze-frame.

However, inside are electronics that record the picture in a wider bandwidth of frequencies. The luminance carrier for Super VHS is placed between 5.4 MHz to 7 MHz range, a deviation of 1.6 MHz. This is opposed to the conventional VHS decks, as shown in Fig. 2-31, that have a 1 MHz luminance carrier deviation between 3.4 and 4.4 MHz. The wider carrier bandwidth, coupled with the shift to the higher frequency range of 5.4 to 7 MHz, provides the Super format with higher resolution—more than 400 horizontal lines as compared to 250 lines in a conventional VHS deck.

Another constituent of Super VHS is separate luminance and chrominance signals. In a conventional VCR, these signals are blended together when recorded on tape and processed by the TV, but are separated before you see the picture. In a Super VHS deck, the luminance and chrominance channels are not mixed, but available at separate output terminals. The separated luminance and chrominance signals reduce crosstalk. The crosstalk appears as a waving herringbone or rippling effect you sometimes see when a tight, regular pattern is displayed in the picture.

Super VHS decks come with three types of video connectors for attaching the machine to your TV: rf, composite video, and Y-C. You get the lowest image sharpness when viewing the picture through the rf terminal, and slightly better quality—about 300 lines of resolution—when watching through the composite video terminal. Your set needs a composite video input on the TV to accept the signal. The full 400-plus lines of video resolution are available only when using the Y-C outputs of the Super VHS deck. Of course, you need a matching set of terminals on your TV.

The maximum resolution of tapes you see on a Super VHS deck really depends

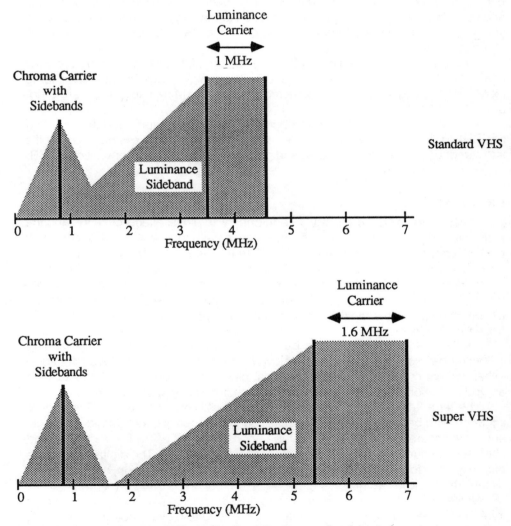

Fig. 2-31. Signal layout for VHS standard and Super recording formats.

57

on the quality of the original signal. If the original is low quality—taped off the air or through cable, for example—the resolution will be the same as with a conventional VCR, even if you use a Super VHS-ready TV. To get full resolution, you must record from a high quality source like a videodisc, play back a pre-recorded Super VHS movie, or make Super VHS tapes with a camcorder.

Because Super VHS uses higher frequencies to record images, tapes made on a Super deck can't be played back on a conventional VCR, even though both machines use VHS format tape cassettes. However, a Super VHS deck *can* play back and record in conventional VHS format, allowing you to share tapes with friends and family and play commercially pre-recorded tapes on your Super deck. But when you try to play a tape recorded with a Super VHS deck on a conventional VCR, all you see is snow.

Tapes made for conventional VCRs are not formulated for the high frequencies used in Super VHS recording; you need an extra high grade cobalt-doped ferric oxide tape to get good pictures. Expectedly, the tape is more expensive than standard grade VHS tape.

A small pin-hole is drilled in the bottom of the Super tape cassette, enabling the deck to distinguish between grades. If the deck senses there is no hole, it shifts into standard VHS mode, recording and playing back tape like a conventional VCR.

DIGITAL EFFECTS

Push the PAUSE button on a VCR equipped with digital effects circuitry, and the picture is sharp and clear—not noisy and jittery as with most home VCRs. Why such crystal clarity? With conventional VCRs, still frames are made by stopping the tape and having the rotating video heads scan the same TV image tracks. The result: jitter, rolling, and noise bars—bands of salt-and-pepper flecks—particularly at the top or bottom of the screen. Digital VCRs use computer random access memory (RAM) chips to store a digitized image of the picture. Thus, you see the still frame frozen in silicon.

Clean still pictures is only one benefit of the newest digital VCRs. Depending on the model, you also get noiseless slow motion and fast motion playback, digital enhancement of the picture during normal playback, unique special effects such as "posterization" and "mosaic," picture-in-picture effects, and more.

The term "digital VCR" is really a misnomer. "Digital effects" VCR is a more accurate term because the digital circuits use video special effect during playback, but it does not record digital data on tape and then play it back. The recording and playback medium is still analog, as with conventional VCR's. The digital circuits are primarily used for special effects, such as freeze-frame, slow motion, and picture-in-picture (also known as PIP).

Digital video effects are achieved by routing the analog playback signal (coming from the tape or through the VCR tuner) to an analog-to-digital (A/D) converter. This chip transforms analog signals into their digital counterparts. The digital information is then processed and stored in dynamic RAM, in much the same way as a memo or spreadsheet is stored in the memory of a personal computer. Digital VCRs have from 64K bytes to about 300K bytes of RAM packed inside them, or about the same as the average desktop computer. Before you see the final picture, the digital data stored in RAM is reconverted using a digital-to-analog (D/A) converter.

Most current digital VCRs use six-bit digitization, so that any picture element, or *pixel,* on the screen is stored as a six-bit binary number in RAM. With six bits, there are a maximum of 64 possible combinations. That's what causes a slight digital look or blockiness during playback. With more bits representing each picture element, there would be more resolution in the final picture, so it would look more uniform.

Once in digital form, the picture data can be manipulated to create unusual special effects. For the posterization and mosaic effects built into some of the digital effects models, the data is read out in different ways.

Some digital effects VCRs have a PIP feature. With PIP, a smaller picture from the VCR tuner or the tape is displayed in one corner of the screen. Controls on the deck let you select which image you want displayed full-size (you only get audio for the full-size picture) and which corner you want the PIP image to appear.

Digitally enhanced slow motion is another common feature to digital effects VCRs. With most decks, the slow motion rate is preset, but on others you can vary the slow motion speed. Many digital effects VCRs have a frame advance feature, which lets you step through still frames one at a time. Frame advance is like slow motion that you control manually. The frame advance can be coupled with picture-in-picture effects to show a sequence of frames on the screen.

Digital circuitry can also be for reducing noise in the playback picture, but as of this writing, only a few models (such as those from NEC) have it. Here is how it works: Each video frame is composed of two fields. The lines from the two fields interlace with one another. During playback, the VCR digitizes the first field. The information from the first field is then averaged with the information from the second field. The averaging acts to cancel out random noise, so the picture looks sharper.

PCM AUDIO

All 8mm decks can record high-fidelity digital sound. The specifications are almost as impressive as those found on audio compact discs: 50 Hz to 15 kHz frequency response, 90 dB signal-to-noise ratio, and 88 dB dynamic range. The analog-to-digital conversion uses 8-bit quantization (by comparison, compact audio discs use superior 16-bit quantization).

The process of digital recording is known as *pulse code modulation,* or PCM. In operation, a series of pulses that represents digital data is recorded on the tape. The pulses are derived by converting the original analog audio signal to digital form. For playback, the pulses are processed in just the reverse, to reconstitute analog audio from the digital.

There are two digital audio modes: video-plus-audio and all-audio (the latter is sometimes called Multi-PCM). In the video-plus-audio mode, the digital sound is recorded with the video picture (the track can also be added after the picture is recorded). In all-audio recording, the deck records only sound. Depending on the

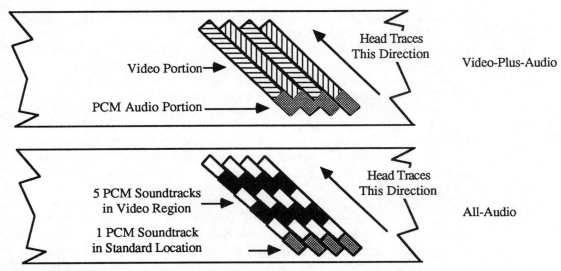

Fig. 2-32. Tracks recorded on 8mm tape in the video-plus-audio and all-audio modes.

recording speed, you can store up to 24 hours of digital sound on a two hour tape (12 hours at SP speed). Figure 2-32 shows the track layout for video-plus-audio and all-audio recording.

Note that in video-plus-audio mode, the PCM audio is placed at the head of each video track. This is accomplished by wrapping the tape farther around the head drum, as depicted in Fig. 2-33. The standard 180 degree wrap is used for video recording and the extra 30 degrees is used for PCM audio recording.

PCM audio is also found on some high-end VHS decks. Most use higher-quality 14-bit quantization.

CONTROLS

Most VCRs have the following typical controls, as illustrated in Fig. 2-34:

- ◆ POWER. Turns power on and off.
- ◆ EJECT. Ejects the cassette from the VCR.
- ◆ PLAY. Places the deck in PLAY mode.
- ◆ FAST-FORWARD. Fast-forwards the tape; no picture on screen.
- ◆ REWIND. Rewinds the tape; no picture on screen.
- ◆ CUE (or Forward Fast-Scan). Scans the tape at 3 × to 30 × normal speed in forward; high-speed picture on screen.
- ◆ REVIEW (or Reverse Fast-Scan). Scans the tape at 3 × to 30 × normal speed in reverse; high-speed picture on screen.
- ◆ PAUSE. Stops the tape during playback; a still frame appears on the screen (the deck stops automatically after about five minutes to prevent excessive tape and video head wear).

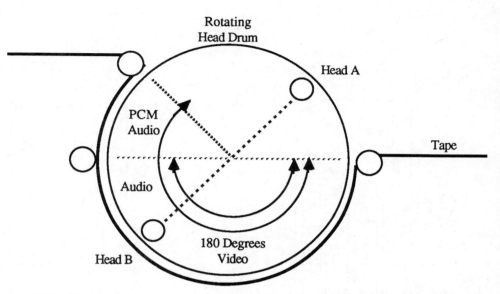

Fig. 2-33. The tape is wrapped around the head in the 8mm system an extra 30 degrees to accommodate the PCM audio track.

- ✦ RECORD. When used with PLAY button, places deck into RECORD mode.
- ✦ AUDIO DUB. Records just audio over previously recorded program (video portion must be previously recorded).
- ✦ VIDEO DUB. Records just video over previously recorded program.
- ✦ TRACKING. Adjusts tracking to reduce or eliminate noise bars in picture (not found on 8mm decks).
- ✦ OTR. One-touch-recording; starts deck in RECORD mode and records for preset period. On many decks, each press of the OTR (or EXPRESS) button increases recording time by 30 minutes.
- ✦ TIME buttons. Sets the record timer.
- ✦ CLOCK buttons. Sets the clock (many decks set time, A.M./P.M., and day of week).
- ✦ CHANNEL 3/4 Output (back of deck). Switches rf modulator output to channel 3 or 4.
- ✦ LINE (with A/V terminals). Switches between rf or direct AUDIO/VIDEO INPUT terminals (might be automatic on some models).

Most VCRs have a counter and counter RESET. The counter is operated by the take-up spindle (either mechanically or electronically) and indicates relative tape position. The counter is reset to ''0000'' by pressing the RESET button.

CONNECTORS

VCRs contain a plethora of connectors located on the front, side, and back of

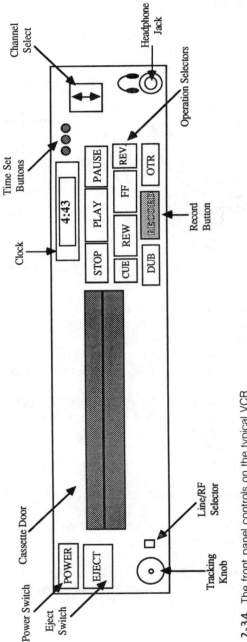

Fig. 2-34. The front panel controls on the typical VCR.

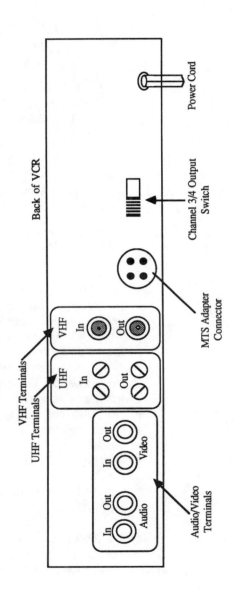

Fig. 2-35. The back panel connectors on the typical VCR.

the deck. Refer to Fig. 2-35 for the typical connectors you're likely to see on the back of a VCR and what they do.

➡ VHF IN. The rf connector for receiving channels 2 through 13. Two sets of connectors can be used—one for 300-ohm twin-lead and the other for 75-ohm coax.

➡ UHF IN. The rf connector for receiving channels 14 through 83. One set of connectors is used for hooking up to 300-ohm twin-lead.

➡ VHF OUT. The rf connector for hooking up to the vhf input of the TV. Two sets of connectors can be used—one for 300-ohm twin-lead and the other for 75-ohm coax.

➡ UHF OUT. The rf connector for hooking up to the uhf input of the TV. One set of connectors is used for hooking up to 300-ohm twin-lead.

➡ VIDEO IN. The video input to the VCR. An RCA phono connector for receiving baseband composite video from another source, such as video disc player, satellite receiver, or another VCR.

➡ VIDEO OUT. The video output from the VCR. An RCA phono connector for hooking up to a TV/monitor, another VCR, etc.

➡ AUDIO IN. The audio input to the VCR. An RCA phono connector for receiving baseband audio from another source, such as video disc player, satellite receiver, or another VCR. Might have two connectors if the VCR is stereo.

➡ AUDIO OUT. The audio output from the VCR. An RCA phono connector for hooking up to a TV/monitor, another VCR, etc. Might have two connectors if the VCR is stereo.

➡ Y-C INPUT. The separate luminance and chrominance video inputs to a Super VHS VCR. A four-prong connector for receiving baseband luminance and chrominance signal from another Super VHS VCR or similar equipped video source.

➡ Y-C OUT. The separate luminance and chrominance video outputs of a Super VHS VCR. A four-prong connector for hooking up to similarly equipped TV/monitor, another Super VHS VCR, etc.

➡ Microphone. The microphone audio input jack, usually accepts a ⅛-inch miniature plug. Might have two jacks if the VCR is stereo.

➡ Headphone. The headphone jack, sized for either ¼-inch standard or ⅛-inch miniature plugs. Stereo VCRs will have a three-way jack for use with stereo headphones.

➡ MTS Decoder. The multi-pin connector for hookup to an outboard MTS stereo decoder. The type of connector will vary depending on make and model.

3

The VCR Environment

Success with your video cassette recorder depends on how well it is integrated with the rest of your video system. A haphazard arrangement will yield inferior results, and you won't be able to enjoy the full capabilities of the deck. Many potential problems can be completely avoided by proper placement and hookup of your VCR.

If your deck is giving you problems, check this chapter first. It provides details on how to properly install, adjust, and use a VCR. Home video isn't necessarily a difficult technology to use; on the contrary, it's quite simple. Nevertheless, a VCR is unlike the trusty TV set, and if you've never owned one before, you might be surprised by the special considerations that you need to keep in mind.

This chapter also details some of the extras you can add to your video system to improve the picture and sound and therefore the enjoyment of your video investment.

UNPACKING

If you have yet to buy or unpack your VCR, here are some quick tips to keep in mind.

Concealed Damage

When unpacking your VCR, be on the lookout for concealed damage. This is breakage that you can't see on the outside of the box but becomes apparent when you take the deck out of its shipping container. If the machine has been damaged, return it immediately to the dealer. Exterior damage is a good indication that internal components are also damaged.

Save the box and all packing material, at least until you've had the deck a month or so. Most mechanical and electrical problems will arise in this time, and you'll need the box to return the deck for warranty repair.

Check Packing List

Most VCRs come with a variety of accessories, including cables, cable adapters, and a remote control. Verify the packing list against the contents of the box to make sure you have everything you are entitled to receive. If you have purchased a camcorder, the unit might come with a cassette adapter (if it is the Compact VHS variety), a battery, battery recharger, and carrying case. Check the manual to be sure that the accessories are present. If any are missing, consult your dealer.

INSTALLATION

Keep these installation tips in mind when adding a VCR to your video system.

Proper Ventilation and the Role of Video Furniture

VCRs generate a certain amount of internal heat, so it is important to avoid obstructing the ventilation slots on the top, sides, back, and bottom of the deck. A machine that gets too hot will perform erratically or can be damaged. Avoid placing the deck on the top of the TV set, a place that normally gets fairly hot after an evening of watching the nighttime soaps.

One way to provide adequate ventilation is with a piece of well-designed video furniture. Video furniture cannot only improve the look of your living room or TV den, but also helps protect your equipment. The least expensive is the roll-around TV and game machine cart. "Bookshelf" furniture has more space for VCRs and accessories, but cost is higher. Most bookshelf units come with wheels or rollers and have adjustable shelves. Large "credenza" furniture can hold a 19-inch or larger TV, VCR, disc player, satellite receiver, tapes, and more. Many have glass fronts and pull-out drawers.

If the furniture you like doesn't have any drawers for small accessories or tapes, consider purchasing a caddy or box that can be filled with these miscellaneous goodies. Be sure the tapes can stand on end (as described later). This lessens the possibility of tape stretching.

When deciding on video furniture, keep in mind that you need to allow at least an inch or two of extra space on all sides—front, top, sides, and rear. This not only allows for easier cable routing and better ventilation, but it lets you install and adjust your equipment with greater ease. Measure each piece of video gear you have, and add at least an inch to every dimension. Remember to include doors that open (front, side, or back access), cables that jut out the back, side, or front, and so forth. Don't rely on the sizes given in the manual; the specifications don't take these important things into consideration.

Next, measure the *inside* of the furniture, and compare these dimensions to the sizes of your equipment. Be sure to measure the distance of the openings in the furniture. These can sometimes be smaller because of trim. Avoid making the installation permanent. Allow yourself the convenience of being able to modify the system.

You should always place the VCR on a level shelf or platform that's free of vibration and shock. Be sure that the shelf can adequately support the deck. If it looks like the shelf might break under the weight of the unit, by all means, find another place to put the machine.

Placement for Remote Control

You should place a VCR that has an infrared remote control in the open, yet out of direct sunlight. If possible, avoid placing it in an audio/video (A/V) rack equipped with a glass front. The glass might disperse the infrared light beam and reduce the effectiveness of the remote. This is especially true if the glass is smoked. You might need to keep the glass front open when operating the VCR with the remote.

Also, be sure that the deck is within range of the remote (usually 20 feet or less), and that you can aim the remote control transmitter in a direct line—more or less—to the VCR. The remote might not work if it is used at angles greater than 30 degrees off to either side of the front panel of the deck, as illustrated in Fig. 3-1.

HOOKUP

Whether or not you plan to invest in a piece of video furniture, you should pay careful attention to how you hook up the pieces of your video system. By using the right kind of cables, connectors, and accessories and wiring everything properly the first time around, you're assured of years of care-free operation. A messy installation only invites aggravation and poor quality video.

Proper Cables a Must

Have you ever taken a ride on a bumpy, country road? After the trip, you probably felt rankled and worn out. And your car! It'll never be the same. Cables are the pathways for the signals generated by your video gear. If those signals have a rough time getting to their destination—that is, the television—the picture and sound quality definitely won't be pleasing.

That's why it's important you use the proper type of cable in your video system. The best cable for video is coaxial, made of a solid center conductor surrounded by a braided outer conductor.

Types of Coaxial Cables

There are actually three major types of coaxial cable (or just "coax") used for video: RG59, RG11, and RG6. The RG6 and RG11 varieties are different only in that the foam surrounding the center conductor is thicker than in RG59. RG6 cable

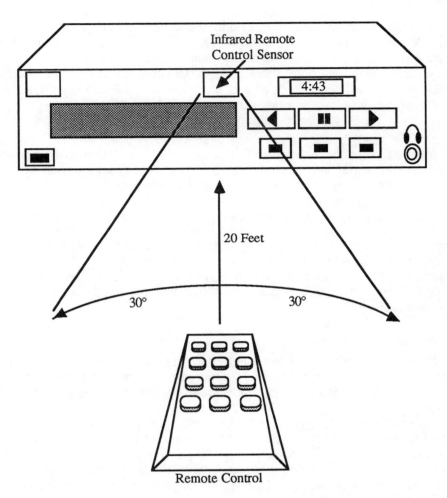

Fig. 3-1. For best results, keep the wireless infrared remote controller within 30 degrees of the axis of the VCR's sensor and no more than 20 feet from the sensor.

is the all-around best because it offers the greatest signal-passing capabilities for all channels, and because it lasts longer under adverse conditions.

Matched Impedances

All three types of TV coax have impedances of 75 ohms. Impedance, which is expressed in ohms, is the measure of opposition an alternating electrical current will encounter when sent down a cable or through a device. It's important that the impedance be matched throughout the entire video system. Ghosts and loss of signal can otherwise occur.

THE VCR ENVIRONMENT

The impedance for the VHF input and output jacks on VCRs and TVs are 75 ohms. So the cables carrying those signals must also be rated at 75 ohms. There are other types of coax cables with different impedances, but they aren't meant for use with video. Stay away from cables designed for CB and amateur radio use (such as RG8 and RG58), because they have an impedance of 52 ohms. If used, they cause a signal imbalance that deteriorates the video signal.

Twinlead

There's another common type of TV hookup cable, called twinlead. It's also used to carry vhf/uhf signals in a video system, but it isn't nearly as good as coax. Why? Coax provides superior rejection of noise and interference because the outside conductor forms a shield. Radio frequency signals radiated from different parts of your video system (and nearby television stations) don't easily penetrate into coax because of the shielding. A video system using twinlead, especially twinlead that's weathered and cracked with age, is typically riddled with ghosts and interference.

Use coax cable with your video system as much as possible. If you have twinlead, outside or inside, consider changing it to coax. If you must mate coax with twinlead (the latter of which has an impedance of 300 ohms), use a matching transformer, discussed below.

The Right Connections

When hooking together coaxial cable, it isn't enough to twist the wires to form a connection. Coax cables need special fittings called "F-connectors," shown in Fig. 3-2, on each end to ensure that the signal passes from cable to equipment without any loss.

The male end of the connector, built into the VCR, TV, or other component, is barrel-shaped with threads around it. The female end of the fitting is attached to the coax, and can be either the thread-on or slip-on type. Care must be taken when attaching an F-connector onto the end of a cable. For information on how to do it, see Appendix E, "Attaching an F-Connector." Or you can buy a ready-made cable with the F-connectors already in place.

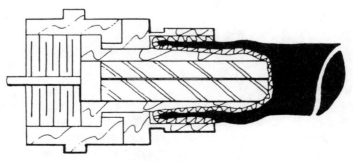

Fig. 3-2. A cutaway view of coaxial cable and F-connector.

Other Types of Cables

Keep in mind that coaxial and twinlead cable are meant for rf signals. For best results, you should use shielded video cable when hooking up components that pass baseband video and audio signals—for example, the VIDEO OUT connector of a VCR to the VIDEO IN connector of a TV/monitor. Again, avoid using shielding cable designed for audio applications. This cable has the wrong impedance, and it isn't sufficiently shielded to reject interference.

You can buy video cable with RCA "phono"-type connectors on each end, like in Fig. 3-3. This is the kind used in consumer video gear that supplies a baseband input or output. Some high-grade baseband cables have a matched impedance of 75 ohms as well as gold-plated connectors. These contribute an extra measure of quality but aren't absolutely necessary.

PARTS FOR PROPER HOOKUP

One of the keys to assembling a top-notch video system is using the proper interconnecting components. There are several major component types, including matching transformers, band splitters, and A-B switches.

Matching Transformer

Matching transformers, also called *baluns,* match the impedance of one type of cable to that of another. They also provide the proper connector or terminal to

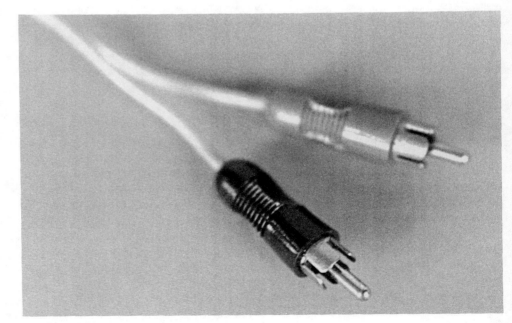

Fig. 3-3. RCA phono connectors.

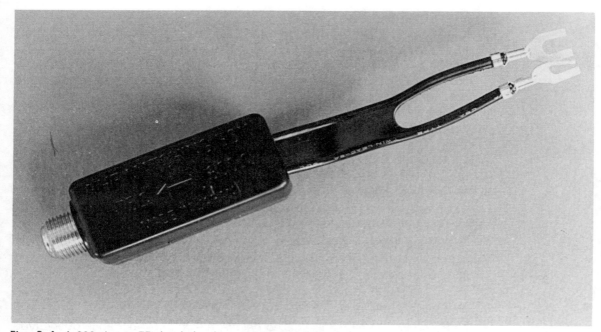

Fig. 3-4. A 300-ohm to 75-ohm balun (or matching transformer).

mate unlike cables together. Figure 3-4 shows a transformer that adapts 75-ohm coax for connection to the VHF terminal on a television set.

Other matching transformers are available for going from 300-ohm twinlead cable or terminals (such as on an outdoor antenna) to 75-ohm coax. Matching transformers are available for both indoor and outdoor use. Outdoor transformers are waterproof to protect them from weathering.

Signal Splitter

Whenever you need to make one vhf/uhf signal go two or more different ways, you need a signal splitter. Splitters are small box-like attachments that have several fittings on them (see Fig. 3-5). Splitters come in three popular styles: Two-way, three-way, and four-way; for splitting a signal two, three, and four ways, respectively.

You should note that a signal splitter divides the *strength* of the signal as well, so you should limit the number of splitters in your system. A four-way splitter divides the strength half again for each output terminal, so you should not use a four-way splitter if you only need a two-way. If you must split the signal so many times that the picture looks grainy or washed out, use an amplified coupler (a splitter with an amplifier built in) or an in-line coaxial amplifier.

In addition, you should avoid leaving one of the outputs of a splitter unused. If you can't fill the vacant spot, attach a terminating resistor to the connector. You might get ghosts without the terminator.

Fig. 3-5. A two-way signal splitter.

Band Splitter

Band splitters take a combined vhf (channels 2 through 13) and uhf (channels 14 through 83) signal and splits it into its separate vhf and uhf components (some even include a separate tap-off for FM radio). Band splitters are used with vhf/uhf antennas and some apartment cable systems. Some band splitters are designed as band combiners, so you can mix vhf and uhf in one cable.

A-B Switch

Another common device you might need is the *A-B switch,* which is used in either of two ways:

1. To route a signal to one of two different destinations.
2. To select between two different input sources.

For example, you can set up an A-B switch (like the one in Fig. 3-6) with your TV so that you can select between either of two VCRs. More elaborate selectors—ones that switch several signals two or more ways—are discussed later in this chapter. There are switchers for both rf and baseband audio/video signals.

THE VCR ENVIRONMENT

Attenuator

You might be lucky enough that the signals from the local TV stations or cable system are too strong. When the signals are overly powerful, you often receive more than one channel at a time. The picture can also look overly dark, have an excessive amount of contrast, or will roll and jitter. These effects are caused by overmodulation distortion.

Attenuators are used to weaken the signals reaching your TV set and VCR. Attenuators can be fixed or variable. A fixed attenuator reduces the signal by some known amount, usually 6 dB. A variable attenuator allows you to adjust the amount of signal reduction, usually from 10 to 20 dB. Variable attenuation is helpful if you receive strong nearby stations and weak, distant stations.

Filters

There are several types of filters available for use with TVs and VCRs that reduce or eliminate interference from other signal sources. One common type is the

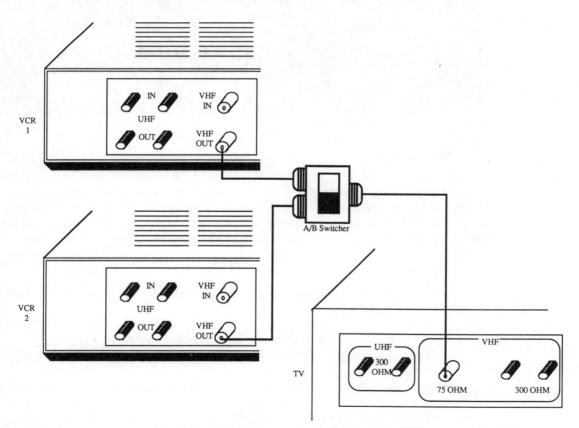

Fig. 3-6. One way to connect an A/B switch to route the signal between two VCRs to one television.

FM filter or "trap," which removes interference caused by nearby FM radio stations. The problem can cause herringbone patterns on your TV when watching channels 5, 6, and 7. In installations where one cable carries both vhf and uhf channels to the TV, FM filters should be used in the vhf line after the vhf and uhf bands have been separated.

Another type of filter, the hi-pass, reduces the crackles and wavy lines caused by interference from local CB and Ham radio sets. Static interference—pops in the sound and flashes on the screen—can be caused by such things as car ignitions and fluorescent lights. A TV interference filter can help cut down the static.

All of the above filters attach between the cable or antenna and the TV. Ac interference filters attach to the power cord of your TV, and help block out annoying static caused by vacuum cleaners, blenders, hair dryers, and other motor-operated appliances.

Signal Amplifier

Signal amplifiers are used when you need to boost the signal coming from the antenna. Amplifiers are especially useful when the antenna is more than 100 feet from the VCR and TV. One type of amplifier attaches to the antenna mast and is powered by a small ac adapter that you connect into the coaxial cable leading to the VCR. The amplifier receives dc power through the center conductor of the cable.

When using such an amplifier, make sure that there are no baluns between the amplifier and the power source. Most baluns block dc electricity, which would render the amplifier useless. Remember that it is always better (and usually cheaper) to replace the antenna with a better model than trying to improve reception by adding an amplifier.

Another type of amplifier, the amplified coupler, is intended for indoor use, and is typically employed to boost the signal so it can be adequately fed to various TVs and VCRs throughout the house. Amplified couplers plug directly into the ac outlet, they don't require a separate power adapter. A sample household wiring diagram, employing an amplified coupler, is shown in Fig. 3-7.

In-line coaxial amplifiers can be placed anywhere in the system and are powered in a similar fashion to mast-mount amplifiers. In-line amplifiers are ideal for boosting the signal when it must pass long distances to a remote TV or VCR.

Grounding Rod

Antennas and masts are made of metal, and because they stick up higher than most other objects around your house, they are susceptible to lightning. For your safety and the safety of your video gear, you should *always* ground the antenna with a grounding rod attached to a static discharge unit on the antenna. With this assembly, much of the dangerous static will pass from the antenna to the earth rather than into your living room.

Installing a grounding rod is simple: Drive the rod firmly into the earth, then connect it using aluminum or copper grounding wire to a static discharge unit

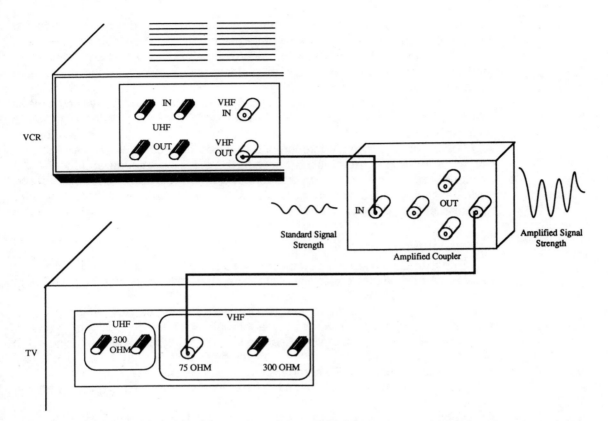

Fig. 3-7. An amplified coupler boosts the signal strength so that it can be sent through long lengths of coaxial cable.

mounted on the antenna. For further protection, you can also install a grounding block anywhere along the antenna cable line feeding your VCR. For best results, locate the block outside where you can easily attach the ground wire to the grounding rod.

Professional Installation Tips

Keep these points in mind when installing your video system.

Plan On Paper. Now that you've got the basics down, what next? When hooking up your system, it's best to place all your video gear—with nothing attached to them— in their proper places. It's a good idea to map out on paper how you'll connect your system before you do any of the actual work. Include all the cables, matching transformers, and splitters you'll need in your drawing. Take the time now to spot any problems. If you look hard enough, you can often find better and less expensive ways to connect your system.

Actual Hookup. When you're satisfied with your hook-up plan, turn off the power to your VCR and TV. Begin connecting the cables and components together starting where the signal is first introduced into your system; beginning just anywhere will lead to confusion if your system is of any size.

Use the shortest cable length whenever possible. If you have extra, DON'T wad it up into a ball or a tight loop. The solid center conductor in coax can be broken if you twist it excessively. Loop it loosely and bind the cable with a plastic tie wrap or sandwich baggie closure. Label each end of the cable for future reference. Connect each cable the same way, making sure you arrange them to avoid tangling.

When hooking everything together, don't staple cables or wires to the inside or down the back of a cabinet or wall. You might puncture the insulation and cause some strange interference, or worse yet, cause a fire or shock hazard. Whenever possible, keep the ac cords separate from cables carrying video signals. Otherwise, you run the risk of video cables picking up interference and noise from the ac wires. If signal cables must cross ac cables, run them at 90 degrees to one another, not parallel. And never place coax cables under carpeting, as walking over them can deform the insulation, which in turn, can create ghosts and snow. If you're running twinlead cable, twist it so that it turns once every 18 inches or so. The twist will help keep out interference.

Testing. When all is finished, turn your TV on and test each portion of your system. Is your VCR working properly? Can it receive all of the channels it's meant to receive? Is the picture crystal clear like its supposed to be? If you're not sure what the picture should look like, hook the cable or antenna wire directly into the TV, bypassing all the other video gear.

Debugging. Problems in installation can occur, but finding the fault doesn't have to be difficult. Go one step at a time. If your VCR isn't receiving a signal, go through your diagram to see if you can find out why. If the problem doesn't lie in your engineering, check the wiring. Substitute cables from paths you know are good. If you must, temporarily reroute or replace cabling to pinpoint the cause.

CHECKOUT

After installation is complete, plug the VCR in a wall socket. Note, as illustrated in Fig. 3-8, the polarization of the plug and socket. Most all electrical devices sold these days have polarized plugs—one prong is slightly larger than the other. They are designed to fit only one way into the wall socket.

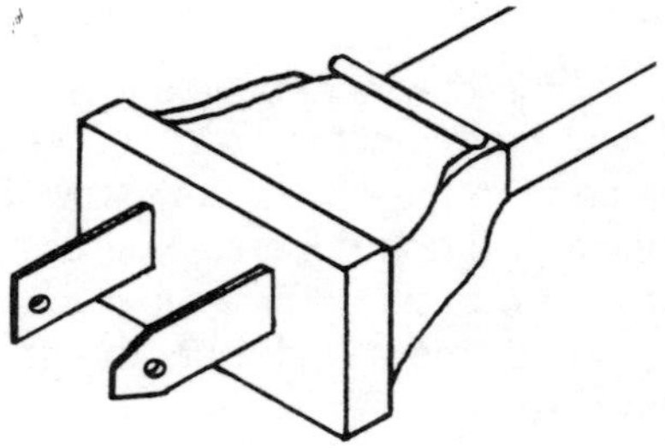

Fig. 3-8. A polarized ac plug, now used on most electrical devices sold in the U.S.

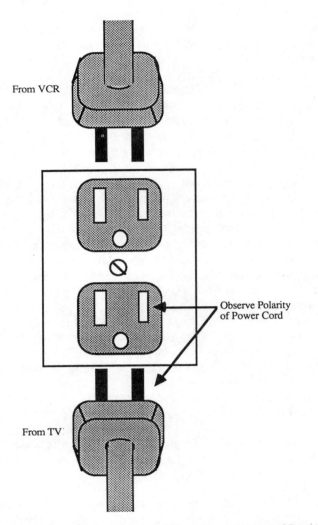

From VCR

Observe Polarity
of Power Cord

From TV

Fig. 3-9. Observe the polarity of the power cords when plugging in the VCR and TV. Reversing the polarity on one device but not the other may introduce electrical hum, which appears as a dark band on the screen or a raspy sound.

If the electrical outlets in your home are not equipped with polarized outlets, be sure to orient the power cable from the VCR in the same direction as the power cable from the TV set, as shown in Fig. 3-10. By retaining the proper polarization for the two components, you eliminate the possibility of a ground loop. Aurally, a ground loop sounds like a low-volume, low-frequency hum. Visually, the loop appears as one or more dark bars that move slowly from top to bottom (or vice versa) of the screen.

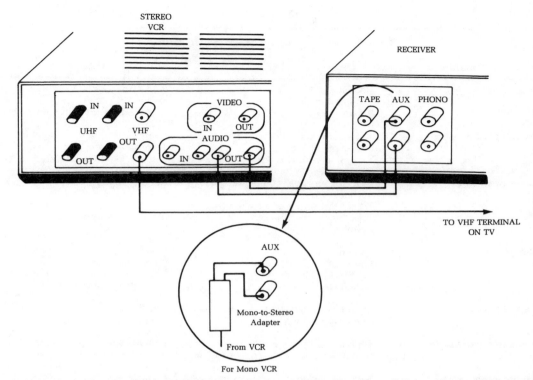

Fig. 3-10. How to hook up your VCR to a stereo receiver or hi-fi amplifier. Use a "Y" (mono-to-stereo) adapter when connecting a mono VCR to the receiver.

Next, turn on the TV (if it isn't already), put a tape into the VCR, and tune the VCR to a broadcast channel. Record several minutes of the broadcast, then play it back. You should see a picture.

Check all of the front panel controls. Press the PAUSE, FAST FORWARD, and REWIND buttons on the deck to make sure everything works. If the VCR has additional features and special effects, consult the operator's manual and try them out.

IF SOMETHING GOES WRONG

Okay. You've connected your VCR to your TV and did everything that you thought you should do, but it doesn't seem to work. Table 3-1 is a quick troubleshooting guide you can use to help correct the mistake. These problems assume a simple cause. For a diagnosis of more serious problems, see the troubleshooting flowcharts in Chapter 9. Bear in mind that many video problems are not caused by the VCR but by the television set, antenna, cable, video tape, and installation. Always suspect a non-VCR-related fault first. Chapter 7 details non-VCR problems and how to solve them.

THE VCR ENVIRONMENT

Table 3-1.

Basic Troubleshooting Guide—VCR Installation		
PROBLEM	**CAUSE**	**REMEDY**
Won't turn on	Not plugged in Switched outlet Blown fuse	Plug into good socket Switch on outlet Replace fuse (usually internal)
Tape won't load or eject	Tape not inserted properly Power off Tape spilled inside deck	Reorient tape and try again Turn power off Remove tape from interior of VCR
No sound or picture	VCR not properly connected to TV Bad or incomplete connection TV set not adjusted or wrong channel Tape blank or damaged Heads dirty	Check cabling Check or replace cabling Check controls on TV Insert known good tape Clean heads
Will not play tape	Tape not inserted properly Power off Tape spilled inside deck Condensation in deck	Reorient tape and try again Turn power off Remove tape from interior of VCR Wait 30 minutes to dry
Picture is snowy	TV set not adjusted or wrong channel Video heads dirty	Check controls on TV Clean heads
VCR does not respond to controls	Tape not properly loaded Switches dirty or electrical problem	Reload tape properly Turn unit off and on again
Remote does not work	Bad or missing batteries in remote Infrared light path blocked	Install fresh batteries Unblock path (including glass)

ACHIEVING OPTIMUM PICTURE AND SOUND

There are a number of ways you can improve the picture and sound to achieve maximum performance from the deck.

Better Picture

If your TV accepts baseband audio and video signals, use the AUDIO and VIDEO OUT connectors on your VCR, not the rf terminal. The reason: when transmitting the signal via the rf terminal, the signal undergoes several unnecessary processing steps, and the sound and picture are degraded because of it. Whenever possible, use the AUDIO and VIDEO baseband terminals.

If you have a Super VHS deck and a television set equipped with Y/C terminals, be sure to use them. The Super-VHS deck is only delivering a percentage of its rated resolution when connected to a TV without the separate Y/C terminals (for more info on Super-VHS, see Chapter 2).

You can greatly enhance the picture quality by adding a signal processor, as discussed in the ''Video Accessories'' section later in this chapter.

Better Sound

By connecting your VCR into your hi-fi system, you greatly improve the quality of the sound. Use a good amplifier and set of speakers. If your VCR is mono, you can use a Y-adapter to branch the one audio line from the VCR to the right and left channels of the hi-fi (see Fig. 3-10). Plug the VCR into the hi-fi terminals marked TAPE or AUX. Never use the PHONO input; the voltage output of the VCR is too high for the PHONO input, and you run the risk of damaging the stereo.

For optimum sound quality, keep these points in mind:

➡ Position the speakers so that they are directed to the central listening point in the room. That point shouldn't be any closer than about eight feet from the speakers. The spot where the sound from the two speakers meet is called the "sweet spot" (Fig. 3-11), and is where the illusion of the "stereo image" is the greatest (assuming you have a stereo VCR).

➡ Position the right and left speakers evenly so that they are the same distance away from the sweet spot. This increases the stereo image.

➡ Speakers almost always sound better when they are positioned on the long wall of the room. If the sound is too "boomy," the speaker placement in that particular room might be causing standing waves. Try a new location.

➡ Place speakers from six to eight feet away from each other. Placing them closer together diminishes the stereo effect; further apart creates a sonic "hole."

➡ Avoid placing the speakers in the corner of the room. This greatly diminishes smooth frequency response.

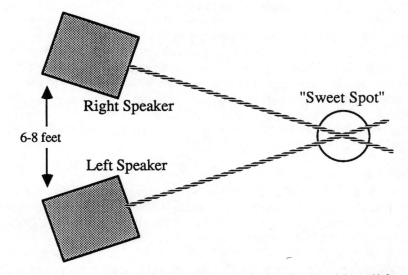

Fig. 3-11. The "sweet spot" is the intersection of the sound waves from the right and left speakers.

THE VCR ENVIRONMENT

➡ Keep speakers off the floor, particularly if the floor is carpeted.

➡ It's best to position the speakers two to three feet from all walls. The closer a speaker is placed to a wall, the more bass it produces. This tip can't always be observed, but try to follow it as much as you can.

➡ If the stereo image is blurred or lacking depth, there might be excessive mid and high frequencies. Dampen these by placing a wall hanging or other soft, absorbing material on the wall between the two speakers.

PROPER TAPE HANDLING

One of the earliest debates to hit the video community was on how to store videotapes. Should they be stored stacked one on top of the other, or on their ends, or backs, or fronts . . . or what?

If you look at the construction of a video tape cassette and take a hint from the old time audiophiles, the answer is fairly clear. It's a generally accepted fact that audio recording tape shouldn't be stored flat (an open reel tape resting on its hub, for example). The strain caused by the reel and the weight of the tape itself can cause warpage. For best results, the tape should be stored upright, so the weight of the reel and tape is distributed more evenly.

Video tapes use internal reels, too, so the same rules of thumb apply. When storing a videotape, place it upright (on edge) either on the "spine" or one of the ends. The spine is where you place the label that describes the contents of the tape, so if you want to be able to easily choose your tapes from the collection in the bookcase, store the cassette on the top or bottom edge.

Technically, it's probably best to store the tape so that the full reel is on the top (the weight of a full reel on the bottom might stretch the tape), but the cassette then tends to be top-heavy. A perfectly acceptable way to store tapes is so the empty reel is on the top. A ratchet mechanism inside the cassette, as shown in Fig. 3-12, prevents undo tape tension so the cassette won't be top-heavy.

By the way, there's no rule that says you must store videotapes in their cardboard or plastic dust jackets. Most seasoned videophiles throw out the tape covers because they get in the way. If you don't use covers, you should at least try to keep the tops of the tapes dust-free. If you live in a dusty area, and don't like constant housekeeping, consider buying a videotape storage cabinet. Make sure it lets you store the tapes the proper way (most do, but some poorly designed cabinets overlook correct tape keeping).

Tape Life

Contrary to what you might have heard, there is no real need to mark the number of times you play your video tapes. A popular myth says that you can use a tape only a certain number of times. This is true, of course, but that number is in the hundreds, and you'll likely never use a tape that much.

Each time you play a tape, you scratch off a little bit of the oxide coating that stores the video and audio signals. At first, the amount isn't much, but after some time—say five or six dozen playings—the results become visually obvious: In the

Fig. 3-12. The ratchet and pawl mechanism in the VHS and Beta cassettes prevents the tape from spilling off the spools. The pawl is released when the cassette is inserted in the VCR.

middle of the picture are flashing streaks of white.

Should you throw the tape out? Not at all. You'll get at least fifty more plays out of the tape, and without the worry of damaging the deck. If, in fact, the tape gets so worn out that it dirties or "clogs" the video heads, it's a simple matter to clean them. When the tape becomes unviewable or repeatedly clogs the heads of your VCR, it's time to toss it.

Damaged Tapes

That takes care of normal tape wear. It's a different matter when the tape is physically damaged. A number of things can hurt a tape. By far, the most common are:

➡ Misthreaded tapes—Where the tape jumps the tape guides inside the deck and gets tangled in the mechanism.

➡ Smashed tapes—Where the door of the cassette closes on tape that hasn't been wound back onto the reels.

➡ Eaten tapes—Where the take-up reel in the VCR doesn't keep up with the tape winding through the transport, and extra tape spills into the works.

Damage doesn't always mean the tape is unwatchable. If the hurt is minor, the tape will still pass through the VCR without harming the heads. There will be a noticeable amount of picture noise for the duration of the damage, however. Severe damage, where the tape is badly mangled, torn, or stretched, means that it's unusable and shouldn't be played.

Videotape *can* be spliced, but it is safe only when the splice is at the very beginning or end of the tape. That's why it's always a good idea to fully rewind a tape before ejecting it; damage can often be repaired at the leader portion without losing the entire tape. Details on how to splice videotape is found in Chapter 5.

For best results, keep your VCR in top working condition. An old, dusty VCR invites tape problems. For example, if the idler wheel gets old and cracked, the take-up reel will slip as the tape is played through the deck. That causes tape spillage—the VCR "eats" the tape, ruining it. This common problem can be avoided by proper maintenance, as described in full detail in Chapter 5, "General Cleaning and Preventative Maintenance."

VIDEO ACCESSORIES

There was a time—not long ago and in a place not far away—when the amateur video enthusiast endured tapes with poor colors, fuzzy images, and rocking and rolling pictures. But that was the limit of the state of the art just five short years ago. Now, however, with more than one quarter of every American household with a VCR, add-on components—little black wonder-boxes—have come along to clear up video's bad image.

But there's a catch. You've probably seen video accessories—image enhancers, color processors, rf switchers, sync stabilizers, and a slew of others—advertised nearly everywhere. To the uninitiated, their purpose isn't always clear. Here's a run-down on the popular types of video accessories. Note that many of the accessory functions are now available in one all-purpose box.

Rf Switcher

The most common video accessory is the rf switcher, so-called because it switches any one of several video signals (more accurately, video signals being sent at radio frequencies) to a number of destinations.

These handy gadgets are rated by their switching matrix; that is, the number of inputs by number of outputs. A typical switcher might have a 4 by 3 matrix (4 × 3), meaning that any of four different signal sources can be switched to any of three different destinations.

Switchers can't be upgraded, so it's important that you buy one that'll fit your present and planned future needs. If you presently have just a VCR along with an outdoor antenna, and want to route the signals to two TVs, in theory you only need a switcher with a 2 × 2 matrix (two inputs by two outputs). But there will likely come the time when you want to upgrade your system—perhaps add a disc player or cable TV input. For most people, a switcher with three or four inputs and two or three outputs is adequate.

The more inputs and outputs an rf switcher has, the more the signals are being split. The strength of rf signals is reduced by one-half each time it is split. A typical signal can be split several times, but excessive signal loss results in a weak, washed-out picture. There is a measurement for signal loss through a switcher, by the way. Look for the unit's *insertion loss,* measured in dB. A loss of 3.5 to 7 dB is considered normal for a 4 × 3 switcher.

To avoid excessive insertion loss, some rf switchers use a built-in signal amplifier. Most amplifiers simply put back what the splitters in the switcher took out, so the end result is an output signal that is the same strength as the input signal.

It's important to keep in mind that rf switchers handle rf signals only. Baseband switchers are also available that switch baseband audio and video signals. Insertion loss is not a consideration when judging baseband switchers.

Rf Modulators

Rf modulators combine a video signal with an audio signal and transmit them over a cable on either TV channel 3 or 4. The better models allow you to select which of the two channels you want; the less expensive models are factory preset.

Image Enhancer

Another major accessory is the image enhancer, which is used to artificially sharpen or add detail to a video picture. Image enhancers work much like the audio equalizers found in home stereo gear: they selectively boost the higher frequency portions of a video signal. The higher frequencies are those that provide sharpness and detail. The more accent given to the high frequency signals, the more detail and sharpness there will be.

Keep in mind, however, that no image enhancer can give back what was never there in the first place. In addition, image enhancers can't put detail back into something that was poorly recorded. The high frequency signals just won't be there to boost.

Image enhancers are best used when making the recording in the first place, particularly when copying a tape. Repeated recordings will wash all the detail out of a tape, so it's helpful to boost the sharpness before the image is re-recorded. The better image enhancers will allow you to control sharpness as well as noise, which is naturally introduced in the enhancing process.

Color Processors

The color processor is another useful video accessory. As explained in Chapter

THE VCR ENVIRONMENT

2, "How VCRs Work," color is transmitted in the television band in such a way that its hue and intensity can be artificially manipulated. To increase or decrease color, the level of the burst signal can be weakened or strengthened. To change color, the phase of the burst signal is changed.

But why change and manipulate colors? There are several reasons. One is that colors can shift, particularly when creating tapes with a color camera. Greens can become blues, reds can become greens and so forth. A color processor can be used to bring the colors back into line. Another reason is that old tapes can fade. Because a color processor can boost the intensity of color far beyond that originally recorded, it's possible to rejuvenate old or poorly recorded tapes with a single twist of a knob.

Old movies broadcast on television are also good contenders for color processors. Because of their age, they are often washed out. Adjust the "chroma gain" to increase the vibrancy of the colors; adjust the "chroma phase" to change the colors (the names of the controls differ from model to model, but most processors use this nomenclature).

Color processors can also be used to vary the overall brightness of the entire picture—color and all. You can bring the scene from normal brightness to pitch black. Separate boxes known as video faders do the same function and can be used to provide special effects to your home tapes. The best faders will include an audio control as well, so both video and audio signals can be faded together.

Video Stabilizer

If you buy or rent commercially produced videotapes, there is one video accessory that you might not be able to get along without: the video stabilizer. Many of the companies that distribute pre-recorded movies change the synchronization signals on the tape slightly to render the video signal uncopyable by a VCR. Most systems omit or invert some or all of the vertical synchronization signals that occur 60 times per second during video transmission.

Unfortunately, many TV sets—particularly the new, all electronic ones—can't handle this change in sync signals. The result: a picture that flips and rolls as if the vertical hold were maladjusted. On some sets, in fact, all that's needed is to adjust the vertical hold and the picture is stable.

But there's the rub. Most of the newer sets lack a vertical hold control or have one inconveniently located on the back or even inside the chassis. The only alternative—apart from not watching copy-protected tapes—is to add a video stabilizer to your VCR.

Sync stabilizers simply put back that which was taken away (more or less). A knob on the front of the stabilizer allows you to fine tune the timing of the sync signals to get the best correction possible. You'll have to readjust the knob each time you watch a new tape, however.

A new form of anti-copying process, called Macrovision, is used on many of the newer movie releases. Macrovision disrupts the automatic gain control circuitry in VCRs, making VCR-to-VCR copying difficult. The result, as seen on the copied tape, is a picture that changes brightness with occasional bright flashes of light. The audio portion is unchanged.

Some television sets are susceptible to the Macrovision encoding scheme, and a number of companies have video stabilizers that attempt to correct for the offending Macrovision signals. The Macrovision process can also impair the picture when you are not copying. Chapter 7, "Troubleshooting and Repairing Non-VCR Problems," addresses this topic.

Stereo Synthesizers/Decoders

Stereo broadcasting is still fairly new. If you have a TV or VCR equipped with an MTS tuner (or a decoder), you can receive broadcasts that are transmitted in stereo. When the program you're watching isn't in stereo, you can still simulate the fuller sound with a stereo synthesizer. The synthesizer hooks up to your hi-fi and you can vary the amount of stereo effect.

Many synthesizers are combo units that also incorporate Dolby "surround" decoding, a stereo process used in movie theaters. With surround audio, you install an additional back (and sometimes front) speaker for added listening dimension. You must have a stereo VCR (hi-fi or linear) and play an encoded pre-recorded tape (such as "Raiders of the Lost Ark"), or watch Dolby-encoded programming from a satellite or cable channel. Surround audio decoders are also available as stand-alone units without the stereo synthesizer circuitry.

4

Tools and Supplies
for VCR Maintenance

The tools and supplies necessary for proper VCR maintenance are not extensive or expensive. Apart from common household tools, you need only a few specialty items to maintain your VCR in good operating condition and to diagnose minor problems. This chapter details these tools and supplies and how to use them.

WORKSPACE AREA

Regular VCR upkeep does not require removal of the machine from its comfortable nest in your video system. As long as you dust the cabinet of the deck regularly, and provide adequate ventilation for keeping the power supply cool, you need only remove the VCR for repair or a preventative maintenance interval.

For minor repair and preventative maintenance, you need a work area that's reasonably free from dust, is well lit, and is comfortable for you. Avoid taking your VCR out in the garage, where there is a greater chance for dust and airborne oil to contaminate its inner workings. A kitchen table or inside work area is ideal.

Before going to work on the deck, lay out a small piece of clean carpeting or heavy fabric over the work table. This will protect the table as well as the machine from scratches and dents. Collect all the tools you'll be using and have them on hand, preferably in a tool box. Special tools and supplies can be stored in an inexpensive fishing tackle box. The tackle box has lots of small compartments for placing the screws and other parts that you remove. If you don't use a box, borrow a cup or bowl from the kitchen to temporarily store parts you remove from the VCR.

For best results, your workspace should be an area where the VCR will not be disturbed if you have to leave it for several hours or several days. The work table should also be one that is off limits or inaccessible to young children, or at least an area that can be easily supervised. Some of the chemicals used to clean VCRs are highly toxic, so you should keep them out of reach of children.

More importantly, there is a risk of electric shock when the top of the VCR is removed. Take every precaution to avoid injury and never leave the deck unattended where curious fingers can touch high voltage wires.

BASIC TOOLS

You'll need a screwdriver and a few other common tools to disassemble the deck. Most VCRs use Phillips screws to contain the cabinet, chassis, and internal components, so be sure you have a Phillips screwdriver handy. Some decks use flathead or hex screws, but these are the exception, not the rule. Determine which tools you need ahead of time and obtain the proper ones. Don't try to make do with the wrong tool. Using a small flathead screwdriver to loosen an allen screw only strips the head of the screw.

If your screwdrivers are not already magnetized, purchase a screwdriver magnetizer at the hardware store. When magnetized, the screwdriver holds on to screws that you remove or re-install into the VCR. The magnetizer lets you magnetize and demagnetize your screwdrivers and other metal tools. Or, you can magnetize your screwdrivers by scraping the blade across the face of a large speaker magnet.

Be sure to save all the screws you remove; they are not as easily replaced as you think. Most VCRs are made in Japan and use hardware with Japanese metric threads, or they use special metal or plastic self-tapping screws. You can't easily find them at hardware stores and they can be expensive if purchased at specialty outlets.

A pair of pliers is also a handy tool to have. The pliers are used to loosen or tighten nuts, grommets, and plastic standoffs. Tweezers or a pair of small long-nosed pliers help you grasp small parts—like screws that have fallen inside the deck! A pair of regular manicuring tweezers is fine, but try to get the kind with the flat, blunt end. Tweezers with a pointed end aren't as useful.

VOLT-OHMMETER

A *volt-ohmmeter* is used to test voltage levels and the impedance of circuits. This moderately priced electronic tool is the basic requirement for intermediate VCR maintenance and repair and is necessary for anything beyond routine cleaning. If you don't already own a volt-ohm meter you should seriously consider buying one. The cost is rather minimal considering the usefulness of the device.

There are many volt-ohm meters (or VOMs) on the market today. For work on VCRs, you don't want a cheap model and you don't need an expensive one. A meter of intermediate quality is sufficient and does the job admirably. The price for such a meter is between $30 and $75 (it tends to be on the low side of this range). Meters are available at Radio Shack and most electronics outlets. Shop around and compare features and prices.

Digital or Analog

There are two general types of VOMs available today: digital and analog. The difference is not that one meter is used on digital circuits and the other on analog circuits. Rather, digital meters employ a numeric display not unlike a digital clock or watch. Analog VOMs use the older fashioned—but still useful—mechanical movement with a needle that points to a set of graduated scales.

Digital VOMs used to cost a great deal more than the analog variety, but the price difference has evened out recently. Digital VOMs, such as the one shown in Fig. 4-1, are fast becoming the standard; in fact, it's hard to find a decent analog meter anymore.

Analog VOMs are traditionally harder to use because you must select the type and range of voltage you are testing, find the proper scale on the meter face, and

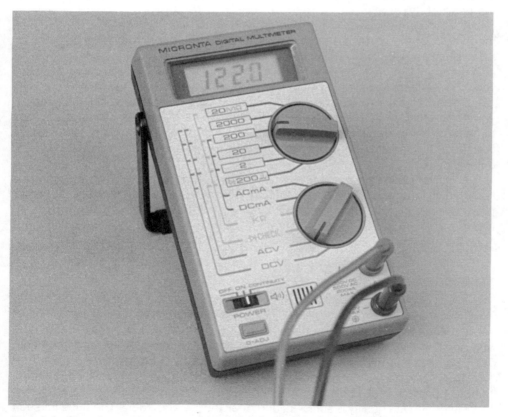

Fig. 4-1. One of several dozen models of digital volt-ohm meters (DVOMs).

then estimate the voltage as the needle swings into action. Digital VOMs, on the other hand, display the voltage in clear numerals, and with a greater precision than most analog meters. Because of their increased popularity and ease of use, this chapter will concentrate on digital VOMs exclusively.

Automatic Ranging

As with analog meters, some digital meters require you to select the range before it can make an accurate measurement. For example, if you are measuring the voltage of a 9-volt transistor battery, you set the range to the setting closest to, but above, 9 volts (with most meters it is the 20- or 50-volt range). Auto-ranging meters don't require you to do this, so they are inherently easier to use. When you want to measure voltage, you set the meter to VOLTS (either AC or DC) and take the measurement. The meter displays the results in the readout panel.

Accuracy

A limited amount of the work you'll do with VCRs requires a meter that's super accurate. A VOM with average accuracy is more than enough. The accuracy of a meter is the minimum amount of error that can occur when making a specific measurement. For example, the meter might be accurate to 2000 volts, ±0.8 percent. A 0.8 percent error at the kinds of voltages used in VCRs—typically 6 to 12 volts dc—is only 0.096 volts.

Digital meters have another kind of accuracy. The number of digits in the display determines the maximum resolution of the measurements. Most digital meters have 3½ digits, so it can display a value as small as .001 (the half digit is a "1" on the left side of the display). Anything less than that is not accurately represented; then again, there's little cause for accuracy higher than this when working on a VCR.

Functions

Digital VOMs vary greatly in the number and type of functions they provide. At the very least, all standard VOMs let you measure ac volts, dc volts, milliamps, and ohms. Some also test capacitance and opens or shorts in discrete components like diodes and transistors.

For most purposes, these additional functions are not necessary, and you need not spend the extra money on a meter that includes them. To make effective measurements, you need to take diodes and transistors out of the circuit to accurately test them. The design of the latest VCRs makes this difficult and inadvisable, even for a seasoned repair technician. (Component failures are usually repaired by service techs by swapping the entire circuit board, not replacing components.)

The maximum range of the meters when measuring volts, milliamps, and resistance also varies. For most applications including VCR troubleshooting, the following maximum ratings are more than adequate:

TOOLS AND SUPPLIES FOR VCR MAINTENANCE

Dc volts	1,000	volts
Ac volts	500	volts
Dc current	200	milliamps
Resistance	2	megohms

Meter Supplies

Most meters come with a pair of test leads—one black and one red—each equipped with a needle-like metal probe. The quality of the test leads is usually minimal, so you might want to purchase a better set. The kind with coiled leads are handy. They stretch out to several feet yet recoil to a manageable length when not in use.

Standard leads are fine for most routine testing, but some measurements might require the use of a clip lead. These have a spring-loaded clip on the end; you can clip the lead in place so your hands are free to do other things. The clips are insulated to prevent short circuits, and you can get clips that attach onto regular test leads.

Meter Safety and Use

Most applications of the meter involve testing low voltage and resistance, both of which are relatively harmless to humans. Sometimes, however, you might need to test high voltages—like the input to a power supply—and careless use of the meter can cause serious bodily harm. Even when you're not actively testing a high voltage circuit, it could be exposed when you remove the cover of the VCR.

If the deck is plugged in while the cover is off (which it will be if you're testing for proper operation) and you're not measuring voltages at the power supply, cover the power supply terminals, if exposed, with a piece of cardboard or insulating plastic.

Proper procedure for meter use involves setting the meter beside the unit under test, making sure it is close enough so that the leads reach the test points inside the deck. Plug in the leads and test the meter operation by first selecting the resistance function setting (use the smallest scale if the meter is not auto-ranging). Touch the leads together, and the meter should read 0 ohms.

If the meter does not respond, check the leads and internal battery and try again. If the display does not read 0 ohms, double-check the range and function settings and adjust the meter to read 0 ohms (not all digital meters have a zero adjust, but most analog meters do).

Once the meter has checked out, select the desired function and range, and apply the leads to the VCR circuits. Usually, the black lead will be connected to ground, and the red lead will be connected to the various test points in the VCR.

NEVER blindly poke around the inside of a VCR in an attempt to get some kind of reading. Apply the test leads only to those portions of the deck that you are familiar with—switch contacts, power contacts, and so forth. If you have a schematic diagram for the VCR, refer to it for the location of the test points.

One safe way to use the meter is to attach a clip on the ground lead and connect it to the chassis or circuit ground. Use one hand to apply the red lead to the various test points; stick the other hand safely in your pocket. With one hand "out of commission,'' you are less likely to receive a shock if you aren't watching what you're doing.

LOGIC PROBE

Meters are typically used for measuring analog signals. Logic probes test for the presence or absence of the low-voltage dc signals (*bits*) that represent digital data. The ''0's'' and ''1's'' are usually electrically defined as 0 volts and 5 volts, respectively (although the actual voltages of the 0 and 1 bits depends entirely on the circuit). You can use a meter to test a logic circuit, but the results aren't always predictable. Further, many logic circuits change states quickly (pulse) and meters cannot track these voltage switches fast enough.

Logic probes, such as the model in Fig. 4-2, are designed to give a visual and (sometimes) aural signal of the logic state of a particular circuit line. One LED on the probe lights up if the logic is 0 (or low), another LED lights up if the logic is 1 (or high). Most probes have a built-in buzzer, which has a different tone for the two logic levels. That way, you don't need to keep glancing at the probe to see the logic level.

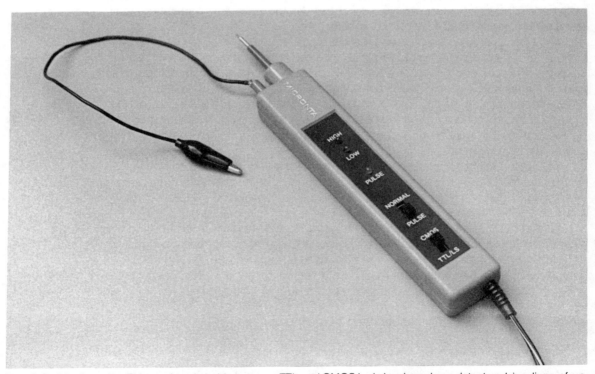

Fig. 4-2. A logic probe. This one is switchable between TTL and CMOS logic levels and can detect pulsing lines of up to 10 MHz.

A third LED or tone might indicate a pulsing signal. A good logic probe can detect that a circuit line is pulsing at speeds of up to 10 MHz, which is more than fast enough for VCR applications. The minimum detectable pulse width (the time the pulse remains at one level) is 50 nanoseconds, again more than sufficient for testing video cassette recorders.

Although logic probes might sound complex, they really are simple devices, and their cost reflects this. You can buy a reasonably good logic probe for under $20. Most probes are not battery operated; rather, they obtain operating voltage from the circuit under test.

Unless you plan in-depth troubleshooting and repair of your VCR, a logic probe is not as important as a VOM. It is a handy tool to have should the need ever arise, but routine maintenance does not require it.

Using a Logic Probe

The same safety precautions apply when using a logic probe as they do when using a meter. When the cover of the deck is removed, potentially dangerous high voltages might be exposed. If you are working close to these voltages, cover them to prevent accidental shock. Logic probes cannot operate with voltages exceeding about 15 volts dc, so if you are unsure of the voltage level of a particular circuit, test it with a meter first to be sure it is safe.

Successful use of the logic probe really requires you to have the circuit schematic to refer to. It's nearly impossible to blindly use the logic probe on a circuit without knowing what you are testing. A single VCR circuit board can contain components for both digital and analog signal processing, and you must know exactly what each component does and how it is used. And because the probe receives its power from the circuit under test, you need to know where to pick off suitable power. You can easily damage the probe—and possibly the circuit under test—if you connect the power leads incorrectly.

To use the probe, connect the probe's power leads to a voltage source on the board, clip the black ground wire to circuit ground, and touch the tip of the probe against a pin of an integrated circuit or the lead of the component. Figure 4-3 shows a probe testing the logic level at an IC pin.

Please note: VCRs often use negative voltages with respect to ground. To the logic probe, there is little difference between connecting the power leads to a positive and ground rail, or connecting the power leads to a negative and ground rail. However, avoid connecting the power leads to a positive *and* negative rail, because the voltage differential might exceed the maximum supply voltage of the probe. For instance, connecting the lead between the +9 and −9 rails will feed 18 volts to the probe, which is higher than its rated operating voltage.

LOGIC PULSER

A handy troubleshooting accessory when working with digital circuits is the *logic pulser*. This device puts out a timed pulse, letting you see the effect of the pulse

Fig. 4-3. Using a logic probe. Care must be exercised to avoid shorting out the pins of ICs and other components. Use probe clips (available at most electronics stores) to anchor the probe to the test point.

on a digital circuit. Normally, you'd use the pulser with a logic probe or an oscilloscope (discussed below). The pulser is switchable between one pulse and continuous pulsing.

Most pulsers obtain their power from the circuit under test. It's important that you remember this. With digital circuits, its generally a bad idea to present an input signal to a device that's greater than the supply voltage for that device. In other words, if a chip is powered by five volts, and you give it a 12-volt pulse, you'll probably ruin the chip. Some circuits work with split (+, −, and ground) power supplies, so be sure you connect the leads of the pulser to the correct power points.

Also be sure that you do not pulse a line that has an output but no input. Some integrated circuits are sensitive to unloaded pulses at their output stages, and improper application of the pulse can destroy the chip.

Making Your Own Pulser

You can make your own pulser out of a 555 timer IC. A suitable schematic is shown in Fig. 4-4. The pulser can be constructed on a small piece of perforated board or universal circuit board. Power for the pulser should be derived from the circuit under test (see discussion below) or it can be powered by a battery that delivers less dc voltage than the circuit you are testing. The 555 chip operates over a wide

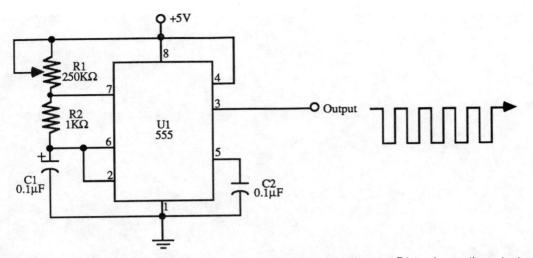

Fig. 4-4. Schematic diagram for the home-made logic pulser. Vary potentiometer R1 to change the output rate.

range of voltages, from about 3 to 15 volts, making it suitable for interfacing with a variety of digital circuits. A finished pulser is shown in Fig. 4-5.

OSCILLOSCOPE

An *oscilloscope* is a pricey tool—good ones start at about $500—and only a small number of electronic hobbyists own one. For really serious work, however, an oscilloscope is an invaluable tool, one that will save you hours of time and frustration. Things you can do with a scope include some of the things you can do with other test equipment, but oscilloscopes do it all in one box and generally with greater precision. Among the many applications of an oscilloscope, you can:

◆ Test dc or ac voltage levels.
◆ Analyze the waveforms of digital and analog circuits.
◆ Determine the operating frequency of digital, analog, and rf circuits.
◆ Test logic levels.
◆ Visually check the timing of a circuit, to see if things are happening in the correct order and at the prescribed time intervals.

The troubleshooting and maintenance procedures covered in this book do not require the use of an oscilloscope, but you'll probably need one if you want to delve deeper in VCR repair. Many service manuals and schematics for VCRs indicate the proper waveform at specific test points (the *waveform* is the visual representation of the electrical signal).

A basic, no-nonsense model is enough, but don't settle for the cheap, single-trace units. A dual-trace (two channel) scope with a 20 to 25 MHz maximum input

94

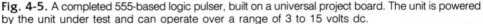

Fig. 4-5. A completed 555-based logic pulser, built on a universal project board. The unit is powered by the unit under test and can operate over a range of 3 to 15 volts dc.

frequency should do the job nicely. The two channels let you monitor two lines at once, so you can easily compare the input signal and output signal at the same time. You do not need a scope with storage or delayed sweep, although if your model has these features, you're sure to find a use for them sooner or later.

Scopes are not particularly easy to use; they have lots of dials and controls that set operation. Thoroughly familiarize yourself with the operation of your oscilloscope before using it. Knowing how to set the time-per-division knob is as important as knowing how to turn the scope on. As usual, exercise caution when using the scope with or near high voltages.

FREQUENCY METER

A *frequency meter* (or *frequency counter*) tests the operating frequency of a circuit, and like the oscilloscope, is not an absolute requirement for intermediate-level VCR maintenance and troubleshooting. Most models, like the one shown in Fig. 4-6, can be used on digital, analog, and rf circuits, for a variety of testing chores—

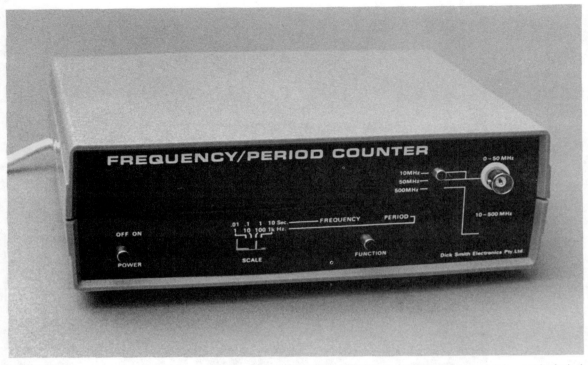

Fig. 4-6. A frequency counter, switch selectable between 10 MHz, 50 MHz, and 500 MHz (a prescaler, not included in this model, is required for 500 MHz operation). This counter was built from a kit.

from making sure the color crystal in the VCR is working properly, to determining the radio frequency of the rf output of the deck. You need only a basic frequency meter—a $100 to $200 investment. You can save some money by building a frequency meter kit.

Frequency meters have an upward operating limit, but it's generally well within the region applicable to VCR applications. A frequency meter with a maximum range of up to 50 MHz is enough. One exception is testing the frequency of the rf output of the VCR, which can extend to 300 MHz and beyond. Higher-priced meters come with or have as options a *prescaler*—a device that extends the useful operating frequency to 500 MHz and above.

INFRARED DETECTOR

VCRs use a number of infrared light sources for normal operation:

➡ Most all VHS VCRs made after 1983 or 1984 use one or two infrared light-emitting diodes as light sources for the end-of-tape sensor. At the beginning and end of the tape is a clear portion (the *leader*). The infrared light passes through the leader, and strikes the sensor. When the sensor detects the presence of light, it knows it has reached the beginning or end of the tape, so it stops. This prevents tape damage during playback as well as during fast-forward and rewind.

➠ Another infrared light source and sensor in the VCR is located under the take-up reel spindle. It detects when the spindle has stopped. If the spindle stops but the VCR is in PLAY or RECORD mode, the deck will automatically shut down after five to 10 seconds to avoid excessive tape spillage.

➠ All but the least expensive VCRs come with a hand-held wireless remote control unit. The control works by transmitting a series of infrared pulses to the deck. When you push a particular button on the controller, a series of invisible pulses is emitted from the controller and received by a sensor in the VCR. The received light is then amplified and decoded, and the VCR acts on your command.

The infrared light sources are almost always light-emitting diodes (LEDs). In VCRs, they operate at about 780 to 880 nanometers of wavelength in the light spectrum, which is just outside normal human vision. Therefore, the ''light'' is actually invisible to your eyes, so there is no way you can visually check the operation of the various infrared LEDs in the deck.

An infrared detector can sense the presence of infrared light and provides you with a visual indication. The sensor is easy to build and inexpensive. It can be used to detect the presence of all types of infrared light, including that from laser diodes in compact disc players and video disc players. That means you can use the sensor to troubleshoot and repair some of your other high-tech gear as well.

Bear in mind that you need not construct the sensor unless you are actively involved in the repair of VCRs or unless you firmly suspect that the LEDs in your deck or remote control unit have gone awry. Infrared LEDs are long-lasting solid-state devices, with an average operating life of 10,000 or more hours, and should outlast most of the mechanical components of the machine. Still, LEDs can break down after only a short period of time, or the circuit providing power to the LED could fail.

If you suspect that an electrical problem is preventing the LEDs from illuminating, you can test the circuit for power with a volt-ohm meter (see above). Position the leads on either side of the LED and take a reading (set the meter to dc volts). Wireless remote control units can also be tested with an ordinary AM radio (see Chapter 5 for details), although this method is not as foolproof because you are testing the operation of the circuits in the remote, not the actual light output.

Many older decks (pre-1982) use an incandescent lamp for the end-of-tape sensor. The lamp, which has a considerably shorter life-span than an LED, might emit visible light, in which case you don't need the infrared detector. However, a number of the earlier decks used an infrared-emitting incandescent lamp. Beware of these; don't needlessly replace the lamp just because you don't see any light output. Use the infrared sensor to make sure that the lamp has indeed burned out.

TOOLS AND SUPPLIES FOR VCR MAINTENANCE

Circuit Description

The infrared light sensor is a simple circuit that employs only three parts: an infrared phototransistor, an LED, and a resistor. The schematic is shown in Fig. 4-7. All the parts are commonly available at Radio Shack and most other electronic parts stores. You need some sort of battery supply to operate the circuit. We've built the circuit with four AA batteries enclosed in a self-contained battery compartment. The compartment also serves as the case for the sensor electronics.

Alternatively, you can use a 9-volt transistor battery mounted in a simple clip holder. The resistor limits the current flowing through the LED, but the 330-ohm value selected is safe for use with supply voltages up to 9 volts. If you use a higher supply voltage, increase the value of the resistor using Ohm's law. Failure to do this can cause the visible LED in the circuit to burn out.

You can determine the value of the resistor by taking the supply voltage and subtracting the voltage drop through the LED (usually 1.2 volts). Divide the result by the current you want flowing through the LED (usually about 20 milliamps).

For example, with a supply voltage of 9 volts, subtracting the drop in the LED makes 7.8 volts. Divide that by .02 (20 milliamps) and you get 390. The exact value of resistor to use, therefore, is 390 ohms. Actually, there is a lot of slack built into this, because LEDs can take higher currents without risk of damage. On a practical level, don't exceed 40 to 45 milliamps.

For best results, solder the components on a small perforated board, or use an 8-pin IC socket. The components fit into the contacts of the socket without soldering, but you might want to solder the leads in place anyway to prevent them from coming out.

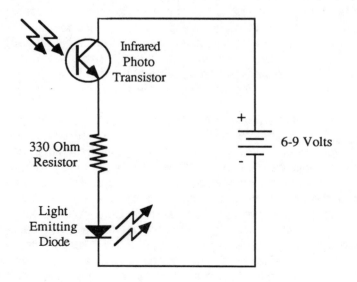

Fig. 4-7. Schematic diagram for the infrared light detector.

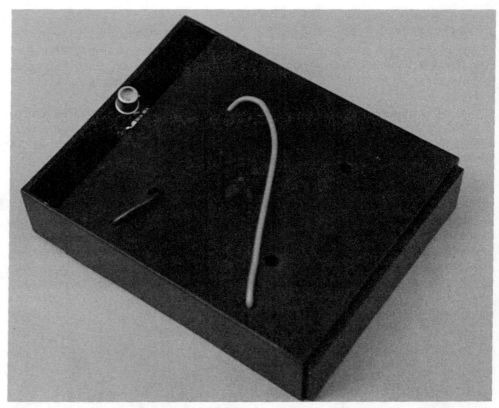

Fig. 4-8. The finished infrared light detector, built into the housing of a 4-AA cell battery pack. The components are secured to an 8-pin IC socket.

Soldering into the socket is a little tricky, because the plastic can melt. An easy way to do it is to fit the component leads into the contacts of the socket and briefly touch the leads with the tip of the soldering iron. Before the plastic of the socket melts, apply the solder. The electrical contact has already been made inside the socket by the contacts; the solder simply serves to anchor the leads in place. Place the assembly in a box, like the battery holder shown in Fig. 4-8, to prevent damage during storage and use. The LED is mounted on the other side of the socket so it can be seen when the sensor is pointed toward the LED under test.

Orientation of both the phototransistor and LED are critical. Installing them backwards will prevent the circuit from working. The phototransistor is marked with a tab or indentation so that you can easily identify the emitter and collector. Refer to the specifications sheet that came with the transistor or consult a transistor guidebook. The plastic rim on the cathode side of the LED is almost always marked with a small indentation. You can install the resistor in any direction. Be sure to connect the power leads in the proper way, too—reversing the leads could destroy the phototransistor.

Using the Sensor

You can test the sensor by connecting the battery and pointing the phototransistor directly into a bright light (the sun and regular light bulbs all have infrared light content). The LED should glow. If it does not, check your work to make sure the components have been wired correctly (of course, be sure the battery is good; test it with your VOM).

Once you're sure the sensor works, you can test the LED in your VCR or remote control unit. You must get the phototransistor within one to four inches from the LED to pick up any light. Figure 4-9 shows the sensor being used to test the light output of the end-of-tape LEDs in a VCR.

TV SET OR MONITOR

A small TV set or monitor is invaluable when troubleshooting and repairing a broken VCR. You can keep the VCR connected to the TV and test the deck as you go through the various troubleshooting procedures. It's a lot easier than hooking the deck back up to your living room set whenever you want to test something.

The TV need not be special, and unless you are troubleshooting problems with the color circuits, the set can be a black and white model. A surplus composite vid-

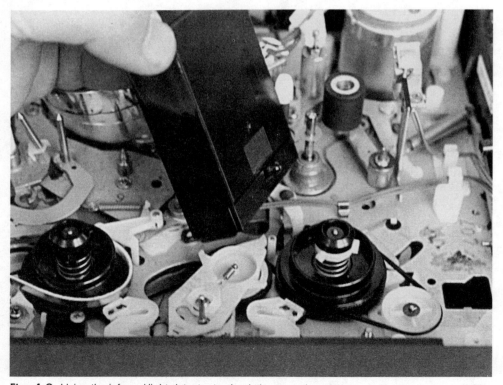

Fig. 4-9. Using the infrared light detector to check the operation of the end-of-tape sensor and LED.

eo monitor can be used with all VCRs that have a separate video output jack (99.99 percent do). Plug the VCR directly into the input of the monitor.

Bear in mind that most monitors lack a speaker, so if you want to hear sound, you need to provide a separate amplifier. A simple transistor battery-powered amplifier (cost: under $15) can be connected to the AUDIO OUT terminals of the VCR. Inexpensive, portable amplifiers use miniature ⅛-plugs. To connect one to your deck, use a phono-to-phono cable, or outfit one end with a phono to ⅛-inch plug adapter.

If the amplifier doesn't have speakers, add one or two small ones. Just about any small speaker will do, as long as they are rated for use with the amplifier. Alternatively, if the amp has a headphone jack, you can use headphones to listen to the audio output. Use caution, however, because unexpected sound coming through the VCR can be deafening if the volume is turned up. When not needed, you should take the headphones off.

A pocket-size TV with a cathode ray tube or liquid crystal display is also ideal for VCR troubleshooting, maintenance, and repair. The TV has a built-in speaker (or earphone jack), and connections for attaching it directly to the VCR. Some pocket TVs have color screens, so you can test the color output of the deck.

ASSORTED SUPPLIES

Unlike the family automobile, the family VCR requires little in the way of oiling and lubricating—if at all. Some cleaning and lubrication supplies might be necessary, however, and any well-equipped maintenance kit will have a little of both.

Spray Cleaner

The exterior cabinet of the deck can be cleaned with a damp sponge. If dirt and grime are a problem, use a mild spray household cleaner, such as Fantastik or 409. Apply the spray to the sponge or cloth, not directly onto the cabinet. Excess can run inside and possibly cause damage.

Cleaner/Degreaser

Freon, the material used as a coolant in air conditioners, is also a solvent. Unlike most petroleum-based solvents, Freon doesn't melt plastics, and the kind of Freon that you can readily buy is non-toxic and non-flammable. It does have a distinctive odor, but it is harmless.

Freon by itself can be used as a basic degreaser and cleaner. You can use it to remove things like grime and dirt in hard to reach places. Freon is available at most any industrial supply house, or if such a business is not nearby, purchase a can of compressed air at a photographic store. Using the can upside down expels the pure Freon propellant.

Freon is often mixed with alcohol to make a more potent cleaner. Use caution with this mixture: it is flammable and toxic. The Freon and alcohol mixture can be used as a general cleaner, a degreaser, even as a video head cleaner. Chapter 5 shows you how to use the mixture to clean video heads and other delicate components.

The Freon/alcohol mixture is available as an all-purpose cleaner/degreaser that you can buy in a spray can. The cleaner is available at Radio Shack and most any electronics supplies store. Figure 4-10 shows a couple of cans that use Freon and alcohol in various mixtures, but they all pretty much do the same thing. The cleaners leave no residue, so you can spray it on and it'll dry with no trace.

The cleaner/degreaser can be used on all the internal components of the VCR, including the printed-circuit board, loading and threading mechanisms, and record/playback heads. Always remember to turn the deck off and let it cool down before spraying. Otherwise, you run the risk of a short circuit. Worse, spraying the cold liquid on warm parts can crack some components. This is especially true of rotary video and VHS hi-fi audio heads. Do not use a spray cleaner on these components if the deck has been recently operated.

Fig. 4-10. Three popular cans of cleaner. The TV Tuner and Control Cleaner (right) includes a lubricant and should not be used for general PC board cleaning or around rubber parts.

Grease and Oil

Though VCRs have numerous mechanical components, most do not need any special lubrication. The reason: the parts are either impregnated with a lubricant (usually a high-viscosity oil) or are made with a material like Teflon that does not require lubrication. Some lubrication might be called for, however, and is especially recommended if the deck has been used in adverse environments (a portable VCR or camcorder would fall into this category) or is more than two or three years old.

Just about any light machine oil can be used for the components that need oiling. The oil should not have anti-rust ingredients. Good candidates are 3-in-1 Oil or most any oil designed for sewing machines. Another good oil is the kind designed for musical instruments. This high-grade oil comes in a handy applicator bottle. Some bottles have a syringe-type needle for applying the oil in hard-to-reach places.

The best oil to use is the kind designed for small machined parts. This oil, which is packed in a small bottle with a syringe applicator, has special penetrating lubricants that ordinary oils lack. You can buy this oil at most industrial supply houses and some camera and electronics stores.

If the oil you have doesn't come with a convenient syringe-type applicator, buy a set of disposal hypodermic needles at the drug store. The needles are usually available behind the counter and can be purchased by adults without a doctor's prescription. Cost is under 50 cents per hypodermic.

To suck the oil up into the hypodermic, open up the can or jar of oil and dip the needle into it. Pull back on the plunger slowly; the oil will sluggishly seep into the hypodermic. To use the syringe, point the end of the needle directly on the spot you want to oil, and push gently on the handle. Apply only a small amount of oil. Exercise care when handling the syringes and always keep the protective caps in place when not in use.

The "grease" should be a high quality industrial lubricant, such as Lubriplate, or a non petroleum-based silicone product. There are a variety of lubricants from which to choose. A light-grade lubricant, such as that used in 35mm cameras, is suitable. You can obtain it from most industrial supply stores and camera repair shops. A small tube goes a long way.

Refer to Chapter 5 for more details on common VCR components that require oiling and lubrication. Note that some mechanical components do not require oil or lubrication and in fact can be harmed by them. Motors fall into this category. It's a safe bet that if there is no sign that oil or lubricant has ever been present on a mechanical part, it does not need it!

Miscellaneous Cleaning Supplies

There are a variety of other supplies that might come in handy when repairing and maintaining a VCR.

➡ Brushes let you dust out dirt. Any good quality painter's or artist's brush will do. Stock a small and a wide brush so you can tackle all jobs.

TOOLS AND SUPPLIES FOR VCR MAINTENANCE

➡ Contact cleaner enables you to clean the electrical contacts in the deck. The cleaner comes in a spray can, but you can apply it by spraying the cleaner onto a brush and whisking the brush against the contacts.

➡ Cotton swabs help you soak up excess oil, lubricant, and cleaner. They are available in quantity at any drug store.

➡ Sponge-tipped swabs are like cotton swabs but leave no lint behind and are ideal for use when cleaning rotary heads. The swabs can be purchased at Radio Shack, most electronics stores, and at the cosmetic counter at the drug store.

➡ Small chunks of untreated, virgin chamois or deerskin are a suitable alternative to sponge-tipped swabs. Purchase the chamois at an auto parts store and cut a portion of it into small one-inch squares.

➡ Orange sticks (from a manicure set) and nail files let you scrape undesirables off circuit boards and electrical contacts.

➡ The eraser on a pencil goes a long way to rub electrical contacts clean, especially ones that have been contaminated by the acid from a leaking battery.

➡ Modeling putty (for plastic models) can be used to mend cracks and chips on the plastic exterior of the VCR cabinet.

➡ Contact cement, white glue, and other common adhesives are excellent for repairing broken plastic and metal parts.

➡ A small magnet makes it easier to pick up screws and ferrous metallic objects that have been accidentally dropped into the deck.

➡ A fluid called "Regrip" helps rejuvenate rubber parts, used in idler wheels and belts. Alcohol should not be used to clean rubber parts, as it can dry them out. Don't use any product that leaves a tacky residue (such as "Non-Slip," available at Radio Shack).

CAPSTAN ROLLER REJUVENATOR LATHE

One of the most common problems of just about any VCR is capstan pinch roller and idler tire wear. The capstan pinch roller is used to provide pressure against the capstan shaft. The shaft and roller pull the tape through the transport mechanism during play and record. If the rubber in the roller becomes old and brittle, playback and recording is impaired.

The idler tire is used to propel the take-up and feed spindles during playback, rewind, and fast-forward. If the tire becomes old, brittle, or dirty, it can slip, and performance of the deck is impaired. With many VCRs, an old idler tire can cause tape spillage, which can ruin the tape and cause jams. Spillage occurs because the take-up reel isn't moving fast enough to wind up the tape that's being fed through the machine during playback.

You can replace the capstan roller and idler tire, but a new part could be outrageously expensive (some VCR manufacturers and retailers might charge as much as $5 to $10 for the parts), or it could be hard to locate. In most cases, you can rejuvenate the rubber of the tire by carefully scraping the outside with a fingernail

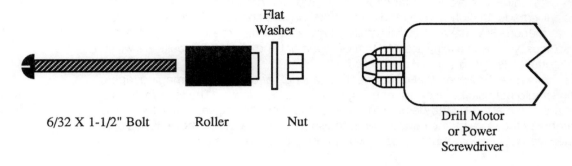

Flat
Washer

6/32 X 1-1/2" Bolt   Roller   Nut   Drill Motor
                                                         or Power
                                                         Screwdriver

To use: Tighten roller on bolt
with nut; attach stem of bolt in
drill motor chuck.

Fig. 4-11. How to use a drill motor to construct a pinch roller rejuvenator lathe. Use a larger or smaller bolt and nut to accommodate the shaft size of the roller or tire you are cleaning.

emery board, as discussed more fully in Chapter 5. You can redress the capstan roller the same way, but a better approach is to use a home-made "lathe," constructed out of a drill motor and spare hardware parts.

Figure 4-11 shows how to construct the makeshift lathe, and how to attach the capstan pinch roller. Use the proper diameter of bolt to accommodate the hole in the roller. Just about any variable-speed drill motor will work (use the drill at a relatively slow speed). A motorized portable screwdriver can also be used, but its chuck might not accept many different bolt sizes. Chapter 5 details the proper use of the lathe.

USING TEST TAPES

Test tapes enable you to put your VCR through a series of diagnostic procedures that examine the deck's ability to successfully play back sound and picture. The tape tests the machine's alignment (the picture will jitter if components are out of alignment) and playback functions. Test tapes are available commercially but can be expensive for the part-time VCR maintenance enthusiast. Most VCR manufacturers (names and addresses are supplied in Appendix A, "Sources") provide test tapes. Write them for a current catalog and price list.

Unless you are repairing VCRs for other people, you don't really need a commercial test tape. There is no need to re-align the internal components of the VCR to meet some industry standard; as long as your VCR plays your old tapes with no problems and can accept pre-recorded tapes, the alignment is perfectly acceptable.

Any tape you made on your deck prior to the occurrence of any problems can be used for testing. It's a good idea, actually, to make a test tape when the deck is brand new. Store the tape in a cool, dry place in case you need it at some later date. Pre-recorded movies are another good candidate for test tapes.

TOOLS AND SUPPLIES FOR VCR MAINTENANCE

The better pre-recorded tapes (of major movies, not off-the-wall art films) are typically made on high-quality tape duplicators that are constantly maintained in top performance. If your VCR successfully plays a tape recorded on one of these machines, you can be fairly assured that its alignment and playback circuits are within acceptable design limits.

Take note, however, that if your deck is "eating" or otherwise damaging tapes, you should not risk the expense of using a pre-recorded movie. The tape might get severely damaged by the deck, and if you are renting the movie, you will need to reimburse the store for the ruined merchandise. Use pre-recorded movie tapes only when you are certain that the deck won't cause physical damage.

General Cleaning and Preventative Maintenance

"An ounce of prevention is worth a pound of cure." The saying applies in many walks of life, and it definitely goes for video cassette recorders. A few well-spent moments keeping your VCR in top-notch condition goes a long way in keeping it healthy. You'll save money in the long run and enjoy your video investment more. A clean VCR is a happy VCR.

In this chapter, you'll learn how to give your VCR a routine preventative maintenance checkup. This includes the deck itself, the deck's remote control, and even your collection of tapes. You'll also learn how to test the machine for proper operation and how to repair moderately damaged tapes.

FREQUENCY OF CHECKUP

The preventative maintenance (PM) interval for your VCR varies, depending on many factors. Under normal use, you'll want to give the deck a PM checkup every 6 to 12 months. By normal use, we mean a VCR operated in an average household environment—not on a factory floor or at the beach.

You might need to perform the PM checkup on a more timely basis if your deck is subjected to environmental extremes or is exposed to heavy doses of dust, dirt, water spray (especially salt water), airborne oil, and sand. A chart of suggested PM intervals for machines that receive light, medium, and heavy use appears at the end of this chapter.

How do you know if your VCR needs a preventative maintenance checkup? Good question. Experience is your only guide, but if you have not yet gained the

GENERAL CLEANING AND PREVENTATIVE MAINTENANCE

experience in how often the deck needs routine maintenance, you can't judge if the service interval is required. As you become acquainted with your video cassette recorder, you will get to know when it needs a checkup. In the meantime, here are some clues that can point you in the right direction:

➡ The outside of the deck is covered with caked-on dirt, dust, or other grime. If the outside is dirty, so is the inside.
➡ The summer has come and gone and you feel your camcorder has seen all the good times with you. Was your camcorder buried in the sand at the all-night beach party? Even if it still plays, it's no guarantee that it's having an easy job at it.
➡ The picture and sound just aren't what they used to be. Something is amiss inside, but not seriously enough to entirely impair playback.
➡ The player makes unusual sounds during tape loading and playback.

A WORD OF CAUTION

It's important to remember that unless you are an authorized repair technician for the brand of deck you own, taking it apart to perform the preventive maintenance procedures outlined in this chapter will probably *void the warranty*. Most VCRs have a warranty period of a year or so (some more, some less), and you might as well take advantage of it. There is usually no reason to perform a PM checkup within the warranty period.

One exception to this is if the machine has been accidentally damaged by water, dirt, or sand, and although it still works, you'd feel better if it were cleaned. Factory warranty service does not cover this type of cleaning, and you'd pay handsomely for it. The cleaning and "repair" cost could well exceed the price you paid for the machine. In this case, there's nothing stopping you from doing the work yourself. If your VCR is subjected to heavy abuse, check the first aid procedures in Chapter 6.

PERSONAL SAFETY

Once again, we are compelled to remind you of the potential dangers that lurk inside your VCR. If yours is an ac-operated home model, removing the cover exposes 110 (or more) volts of alternating current. Given the right circumstances, this current can kill you. Remember to unplug the unit at all times when you are not actually testing its operation. Unless otherwise specified, all maintenance and repair procedures are made with the VCR off and unplugged from the wall socket. Unplugging the unit is an important step. Even with the power switch off, electricity is still present at the terminals near the unit's power supply unless the VCR is unplugged.

Warning signs that appear on the back of ac-operated VCRs (and in their manuals) say there are potentially dangerous voltages inside. When you see the triangle and lightning bolt sign, as shown in Fig. 5-1, exercise caution. The other

Fig. 5-1. Warning signs that appear on the back or bottom of the VCR.

warning sign, the triangle and exclamation point, also shown in the figure, says that you should refer to the owner's manual before operating the equipment.

Warning: Do not perform the steps in this chapter unless you feel confident in your ability and have observed all safety precautions. Take your time and think about every step. If you don't feel you have time for the general cleaning and maintenance procedures, put it off until another day when you have more opportunity. Rushing things will surely lead to disaster.

GENERAL MAINTENANCE

Before you go poking around inside the VCR, first take a close look on the outside. You can perform these preventative maintenance steps on a more regular basis because they do not entail removing the cover of the player.

Manuals of Operation

An important measure in your efforts to minimize VCR troubles is to read the instruction manuals that came with your equipment. This might sound a bit obvious, but you'd be surprised how many video enthusiasts never take the time to thoroughly read the operating manuals. The manufacturers of the equipment include the manuals so you can get the most out of your expensive purchase. Keep them at hand, and refer to them whenever you have a question.

Exterior Dust and Dirt

On the top of your system maintenance list should be routine external cleaning. Every few days, take a dry cloth and wipe each component of your video system.

GENERAL CLEANING AND PREVENTATIVE MAINTENANCE

Don't use dusting sprays; these actually attract dust, luring dirt back onto your equipment. Use a soft, sable painter's brush for those hard to reach places. Be sure to clean the ventilation slots as these are favorite hiding places for dust.

If you need to get rid of stubborn grime, apply a light spray of regular household cleaner onto a clean rag; then wipe with the rag. Never apply the spray directly onto the deck, as the excess can run inside. Never apply a petroleum- or acetone-based solvent cleaner as it might remove paint or melt the exterior plastic parts. Some plastics, when in contact with solvents, let off highly toxic fumes.

Cables

The cables and connectors in your video system can make your VCR sneeze and wheeze too, so make sure they're on tight and that none of them have become damaged. Many owners of VCRs have a tangle of cables behind their TV, and just one bad wire or connector in the bunch can cause grief.

First of importance is coaxial cable maintenance. Coax cable consists of a solid center conductor surrounded by a heavy plastic called the *dielectric.* Around this is a mesh of braid that not only acts as a second conductor but as shielding against external interference. The entire cable is insulated with plastic insulation on the outside.

Coax cannot cope well with torture, yet it's often subjected to twisting, tugging, kinking, and tight looping. If you've got excess, wind it loosely; never bundle it up like extra speaker wire. Be certain that none of your coax is in foot traffic. That includes routing it under carpets. Walking over coax can deform the dielectric, creating ghosts and loss of signal. Tripping over the cable can tug on the connectors, or worse yet, yank your equipment onto the floor.

While inspecting the cabling, make sure to remove any kinks that might cause signal loss and ghosting. Look closely at the connectors attached to the ends of the cable to make sure they are on tight and that the center conductor has not been broken or bent (see Fig. 5-2).

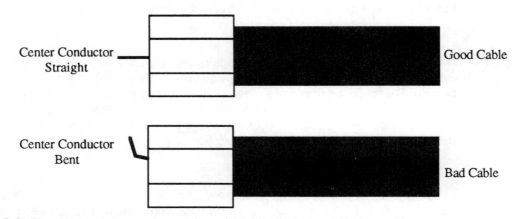

Fig. 5-2. The center conductor on a good coaxial cable should be straight and shiny.

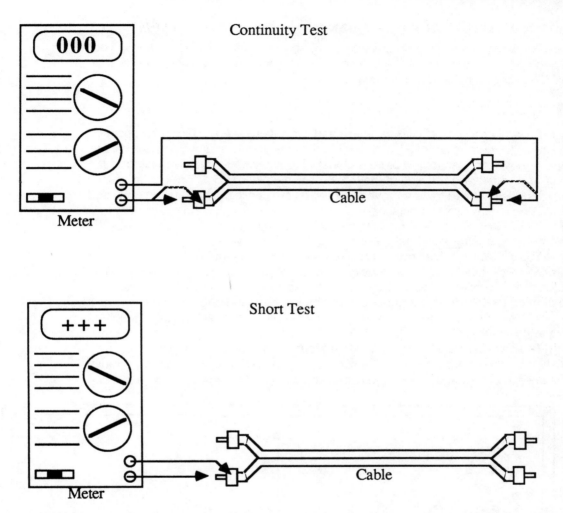

Fig. 5-3. Use a meter to test the condition of cables in your video system. You can test for continuity and shorts by dialing the meter to the ohms (or resistance) setting.

If you suspect that a cable might be bad or causing interference and hum, try a replacement to see if the problem is solved. Alternatively, you can use a volt-ohm meter, as shown in Fig. 5-3, to test the continuity of the cable. Attach the test leads to the center conductor of both ends and take a reading. The meter should read 0 ohms. Do the same for the outer conductor.

Check also that the cable has not shorted by applying the test leads to the inner and outer conductors on one end of the cable (be sure not to hold onto the metal probes of the test leads or you might get false readings). The measurement should read infinite ohms (open circuit).

111

GENERAL CLEANING AND PREVENTATIVE MAINTENANCE

INTERNAL PREVENTATIVE MAINTENANCE

To gain access inside the deck, it should be removed from its location in your video system. Disassemble the VCR only in an open, well illuminated work area, as described in more detail in Chapter 4.

Disassembly

Disassembly of most VCRs is simple and straightforward. The typical ac-operated home unit is composed of a top and bottom piece; you loosen two or more screws and remove the top portion of the cabinet. The screws are usually located at the rear and sometimes the side of the deck, as shown in Fig. 5-4. The bottom panel, which does not need to be removed unless you are servicing the capstan drive and threading mechanisms, is detached by removing the screws on the bottom (see Fig. 5-5). Note that in some early model Sony and Zenith Beta decks, you must first remove the tracking control knob before removing the front cover.

You should not loosen every screw you see, especially those on the bottom center portion of the cabinet, because they might be anchoring some internal components. Loosen one screw at a time and attempt to remove the cover (or at least see if it is coming loose). Run your fingers along the edges of the cabinet to see where it is still catching. If a screw seems to hold an internal component in place, retighten it and try another.

The cabinet screws on some VCRs are marked with an arrow or paint to indicate that they should be removed. If the screws on your VCR are so identified, remove the ones that are marked only. Leave the others. Top-loading VCRs require that you

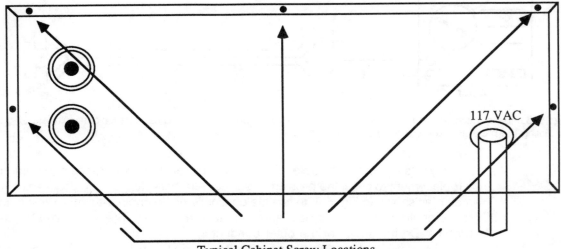
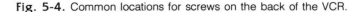

Typical Cabinet Screw Locations

Fig. 5-4. Common locations for screws on the back of the VCR.

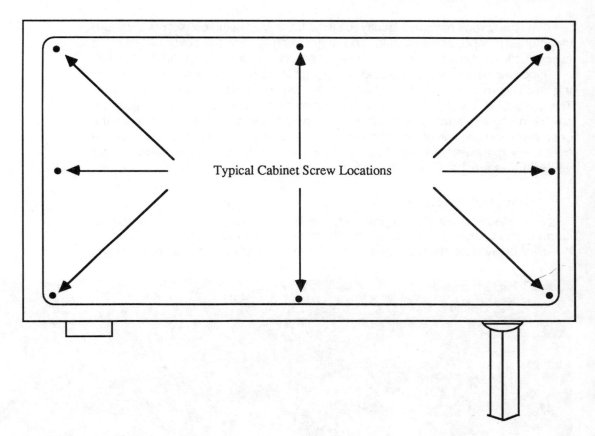

Fig. 5-5. Common locations for screws on the bottom of the VCR.

remove the top plate before removing the cabinet cover. The top plate is held in place by two screws.

Save the screws in a cup or container (or your fishing tackle box). If the screws are different sizes (they probably will be), note where they came from for easier reassembly. Keep washers, grommets, spacers, and other hardware with their related screws. The additional hardware has been used for a reason and should not be omitted when the deck is reassembled.

If necessary, write a list of the parts you remove, and the order in which they were removed. If a certain screw has a rubber washer, a metal washer, and a locking washer on it, write down the order in which they are assembled on the screw. Carefully inspect the parts you remove to determine if they are aligned or oriented in a certain way. Indicate the alignment or orientation on your sheet so you are sure to duplicate it when you put the machine back together again.

GENERAL CLEANING AND PREVENTATIVE MAINTENANCE

With the screws removed, gently lift off the top cover, being careful not to disturb the internal components or wiring. You might need to slide the cover back before taking it off, as shown in Fig. 5-6, if the cover tucks under a lip in the front panel.

Once the cover is off, you should not attempt to disassemble anything else inside the machine, except where noted below. This is especially true of front-loading VCRs. The cassette-lift mechanism can be difficult to reassemble if you don't have the service manual to refer to. In addition, loosening the wrong screws can skew the orientation and position of critical components. This might require a service manual (which is expensive) and a special alignment jig (which you cannot easily get for yourself) to correct the mistake.

About Static Electricity

Static electricity build-up, in your body, can harm the integrated circuits and other components on the main circuit board. Therefore, do not touch any part unless you have first drained the static electricity from your body. If you plan to handle the circuit

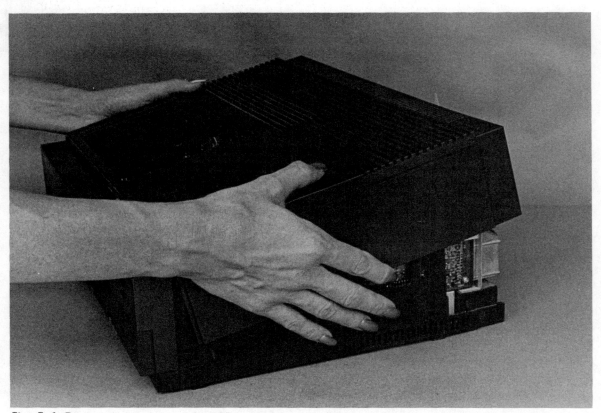

Fig. 5-6. Remove the top cover of the VCR by lifting from the back and sliding the top out from under the lip in the front panel.

Fig. 5-7. Top view of the VCR with the cover removed.

boards, use a grounding strap around your wrist. The strap is available at most electronics supply stores. Follow the instructions supplied with it for proper use.

Preliminary Inspection

Take a few moments to acquaint yourself with the inside of the deck. Note the position of the video head drum, the loading and threading mechanisms, the main circuit board, the auxiliary boards for the front panel, the indicator, the tuner, the video section, the routing of the cables, and the incoming power supply lines. Avoid touching anything, especially the video head drum, unless you absolutely must. Figures 5-7 through 5-12 show the innards of a typical VCR. Figure 5-13 shows the view of the bottom of the VCR, illustrating the capstan drive mechanics and the threading gears and associated mechanism.

If you like, draw a picture of the insides of your VCR, and note where things belong and how they work. When you are well acquainted with your VCR, you will be able to service it better and more confidently. Note especially the location and appearance of all the belts, tires, pulleys, fuses, and other mechanical parts. You

Fig. 5-8. The video head drum.

may want to slip a tape into the deck and watch it load and thread. Note the action of all the parts.

One word of caution here: Avoid operating the VCR in any position but upright. Don't turn the VCR over or on its side, especially during threading. The tape can spill out and become jammed in the works. Operating the VCR in an abnormal position also poses greater stresses on the motors and mechanical parts, which could lead to premature failure.

While the top (and/or bottom) of the deck is off, be particularly wary of the power supply board or power supply section on the main printed-circuit board (PCB). The filtering capacitors (which usually look like tall cans) retain some current even when the player is turned off and unplugged. Avoid touching the leads of the capacitors and make sure that you do not short the leads together. You can damage the capacitor and components on the board.

Front-loading decks obviously have more complex loading mechanisms than their top-load cousins. There is more to look at in a front-load machine and consequently more points that might need preventative care.

Bear in mind that there is little, if any, adjustment or calibration that you can—or should—do once inside the deck. Potentiometers on the circuit boards are factory set and require an oscilloscope or special test circuitry for proper adjustment. The nature of VCRs inherently leaves few internal controls you can safely tweak.

In addition, the latest models are designed with few alignment points. Older models—those manufactured before 1981 or 1982, have service points for fine-tuning the take-up and supply reel torque, pinch roller pressure, back-tension, etc. Today, VCRs are designed with engineering tolerances built-in—these adjustments are rarely found and no longer need to be made.

Tires, Pinch Rollers, and Belts

Visually check the tires and belts used in the reel spindles, the tape loading mechanisms (gears are sometimes used here) idler tires, and capstan pinch roller. If worn or cracked, they should be cleaned or replaced.

To examine the rubber piece, you might need to stretch or twist it, as you would

Fig. 5-9. The transport mechanism.

Fig. 5-10. The main printed circuit board (underside of deck on this model).

an automotive fan belt. Look for a glazed appearance on the rubber; that indicates wear. Cracking is a sign that the rubber is brittle and dry, and is no longer doing its job. It must be replaced. An oily film on the rubber can be an indication that the VCR was excessively lubricated, and that oil or grease from the mechanical sections of the deck have migrated to other components. The oil must be removed to restore proper traction. Healthy rubber will have a suppleness to it. With experience, you will learn to tell the difference between strong, healthy rubber parts, and weak, sickly ones.

Idler tires and pinch rollers can develop flat spots, especially if the VCR has not been used for some time. You can visually inspect for flattening, and if it is only minor, rejuvenation of the rubber—as discussed below—can repair the damage. If the tire or roller is excessively flat, it must be replaced.

You can replace belts yourself. Obtain the service literature for your player from the manufacturer, identify the part by its part number, and order it from the manufacturer. Belts and idler tires (you replace the tire, not the complete mechanism) can often be ordered simply by measuring the rubber's diameter and size. Dentist's picks, which you can buy at most electronics and surplus outlets, let you more easily remove and install belts that are hard to reach. Figure 5-14 shows a typical belt configuration as used in the capstan drive mechanism. Note that the capstan motor drives several sections of the deck.

Rubber rollers and belts that are slipping can be cleaned with a special cleaner called Regrip. The non-slip cleaner is available at many electronic supply stores. Apply it to the belt or roller with a cotton swap or sponge applicator, as shown in Fig. 5-15. Wait a few minutes, then vigorously wipe it off with a cotton swab. Use a disposable fingernail file (emery board) to "sand" the old, top layer. Scrape the flat of the board across the rubber in even, gentle strokes. Don't do more than one stroke at any one spot, or you might make the tire or roller flat.

You can achieve better cleaning results of pinch rollers (and similar parts) by using the home-made motorized "lathe" originally discussed in Chapter 4. The lathe in operation is shown in Fig. 5-16. To use the lathe, mount the pinch roller on the

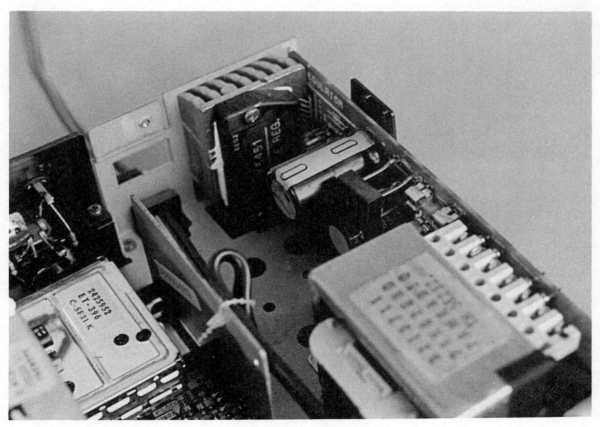

Fig. 5-11. The power supply section (the transformer is in the foreground).

Fig. 5-12. The rf section. The enclosed module is the rf modulator.

bolt. By keeping several sizes of bolts and nuts handy, you can accommodate the shaft sizes of a variety of tires and capstans. Tighten the nuts so that the roller does not spin on the shaft of the bolt. Do not over-tighten, because you might break the pinch roller or deform the rubber.

Rest the drill motor or motorized screwdriver on a flat, hard surface, and turn it on. Reduce the speed to a moderate 100 to 300 rpm. Lightly rub an emery board against the rubber of the roller while it is spinning. Do not press hard; make sure that you run the board evenly across the surface of the rubber, as shown in Fig. 5-17. Periodically stop the lathe and inspect the part. Although the exact diameter of the roller is not crucial, you should not remove more than a thin layer of rubber.

Avoid the use of solvent-based cleaners on rubber parts, as these chemicals can prematurely dry up the rubber, or worse, melt it. Never apply a lubricant of any kind to a tire, roller, or belt, as this will defeat its purpose. If the rubber squeaks, it's an indication that something is not properly aligned, or that the rubber needs cleaning or replacement.

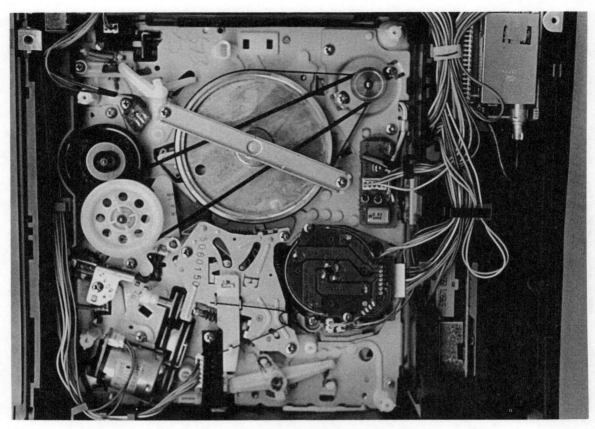

Fig. 5-13. Bottom view of the transport mechanism, showing the capstan flywheel and belts.

Fig. 5-14. The typical arrangement of belt between capstan motor and the pulley that drives the capstan flywheel and capstan shaft.

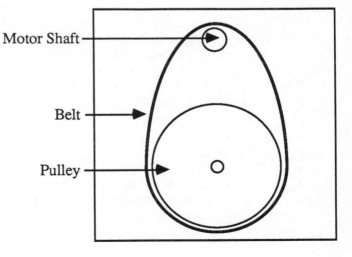

Motor Shaft

Belt

Pulley

Fig. 5-15. Clean the idler tire with a swab soaked in cleaner or rubber rejuvenator.

Tires, belts, and rollers press against plastic and metal parts. After years of use, rubber can cake up on these parts, causing a loss of traction. While you are cleaning the idler tires, belts, and pinch roller, clean the surfaces these parts come into contact with. Use alcohol or a non-petroleum based solvent to clean the part. Caked-on rubber that refuses to come off can be removed with an orange stick, borrowed from a manicure set. Pay particular attention to:

- ✦ Capstan shaft.
- ✦ Reel spindle rims.
- ✦ Idler tire driver.
- ✦ Capstan motor pulley (underside of deck).
- ✦ Capstan shaft flywheel (underside of deck).
- ✦ Threading/loading pulleys.

Fig. 5-16. The lathe being used to restore the surface of a pinch roller.

Brakes and Torque Limiters

The supply and take-up reels of VCRs are outfitted with brakes that prevent free-movement during normal recording and playback. Some VCRs use a number of different brakes: a band-shaped brake is used on the supply reel to compensate for tape slack; a brake on each reel stops the spindles from turning after fast-

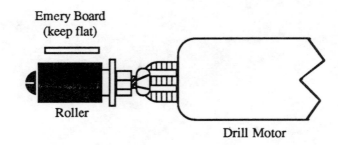

Fig. 5-17. Be sure to keep the emery board flat against the surface of the roller.

forwarding or rewinding; brake pads on both spindles limit the torque of the reels, preventing undo tape tension. A closeup of the brake system in a VHS deck is shown in Fig. 5-18.

Inspect the condition of the brake pads. The band-brake will use a strip of felt—be sure the felt hasn't worn off. The other brakes will use rubber as the brake pads. Inspect the rubber and use Re-grip to rejuvenate it, if necessary. Clean the contact surfaces of the reel spindles with alcohol to get rid of any caked-on deposits.

Topical Cleaning

Dust, dirt, and nicotine from cigarette smoke can accumulate in the interior of the VCR and should be removed. Use a soft brush to wipe away excess dust and dirt. If there is a lot of sediment, use a small hobby vacuum cleaner.

Remaining dust and other junk can be blown out of the insides of the deck.

Fig. 5-18. The brake on the supply reel spindle.

Fig. 5-19. Spraying the main PCB with a non-residue cleaner/degreaser.

Purchase a small can of compressed air at a photographic shop (about $3) and liberally squirt it inside the machine. That should free just about everything that shouldn't be there. Remember to keep the can upright, as the propellant might drip out. As much as possible, position the air flow so that debris is blown *out* of the VCR, not deeper inside!

The use of household cleaning sprays should be avoided, as these not only can leave a residue that impairs the proper operation of the deck, but they are water-based and might short out the circuitry. Do not use a cleaner that contains a solvent of any kind, and DO NOT clean the machine with an oiless lubricant such as WD-40. This spray is non-conductive and will cause the deck to fail completely.

You can clean nicotine sediments and caked-on dirt using a cleaner recommended for application on mechanical and electronic components. Figure 5-19 shows such a cleaner being used on the main PCB (apply the cleaner only when the deck has cooled down after use). Some cleaners have a built-in degreaser that disperses oil build up; these are fine for use in your VCR, as long as they are de-

GENERAL CLEANING AND PREVENTATIVE MAINTENANCE

signed for direct application onto electronic parts. Whatever cleaner you use, make sure it leaves no residue after drying (which usually takes less than 15 seconds). If the cleaner leaves a noticeable residue, it is unsuitable for use inside your deck.

The cleaner is available at most electronics stores and usually comes in spray form. The spray comes out fast, so you can use the pressure to dislodge stubborn grime. Most cans come with an extension spray nozzle tube. Attach the tube to the spray button, and squirt into hard-to-reach places.

Prior to cleaning, be absolutely sure the VCR is unplugged from the wall outlet. Never use the spray where there is live current. You could receive a shock or start a fire. After cleaning, wait at least three minutes for all the cleaner to evaporate before applying power to the deck.

Electrical Contact Cleaning

It's rare that the electrical contacts inside the deck will need cleaning. However, exposure to outside elements can quickly oxidize plated electrical contacts (now you know why you should never place your VCR near an open window where it might be exposed to the damp outside air). The contacts can also be contaminated by an accumulation of cigarette smoke.

If you see a heavy amount of oxidation or nicotine, carefully disconnect the wire leading to the contact and clean the contact with a commercially available contact cleaner. Clean the exposed part of the wire in a similar fashion. If there is a heavy build-up of oxidation, scrape it off with an orange stick or nail file, and then use the cleaner.

Most contact cleaner is in spray form, and the spray might squirt to an overly large area. If you don't want this, spray the cleaner into a cotton ball or swab, then use the cotton as an applicator. Be sure to leave no lint from the cotton on the contacts. If cotton threads are a problem, purchase some sponge-tipped applicators. Most electronics and many record stores sell them. For very hard-to-reach places, use the extension nozzle tube.

Some circuit boards in the VCR will be connected to the main board by an edge-card connector. If you remove the circuit board for servicing or cleaning, be sure that you don't touch the plated "lands" on the board or the connector pins on the main PCB. Not only is there a chance of damaging a component with a shock of static electricity from your body, the oils in your skin will act as a corrosive acid, diminishing the effectiveness of the electrical contact.

VIDEO HEADS

VCRs use spinning magnetic heads to record and play back video information to and from the tape (VHS hi-fi decks also use rotating audio heads, and a few VCR models have rotating flying erase heads). The assembly and the way the tape wraps around the video head drum is shown in Fig. 5-20. Like any other magnetic head, the video heads will eventually become dirty (or "clogged") due to an accumulation

Fig. 5-20. The tape wrapped loosely around the video drum shows the path of the tape through the mechanism of the VCR.

of oxide coating shed from the tapes. The heads can also become dirty by direct contact with dirt, ash, grime, or other contaminates on the tape surface.

Dirty heads can also create excessive dropout. This looks like salt-and-pepper flecks on the screen. Depending on the amount of dropout, the flecks can be anything from a light snow to a heavy blizzard. Routine and regular cleaning of video heads will ensure carefree operation and optimum performance.

The "best" method of cleaning video heads is a debatable question and is one that only you can decide for yourself. Covered here are the two most accepted methods and you may decide which one suits you the most. You might even want to alternate between the two methods, as each one has its own distinct advantages.

Cleaning Cassettes

By far, the easiest and fastest way to clean video heads is by using a commercially manufactured cleaning cassette, like the one pictured in Fig. 5-21. Use is simple: just pop the cassette into the machine, PLAY it for the recommended time (usually less than 20 seconds), and put it away until the next time. A few of the cleaning cassettes display a message on the screen when cleaning is complete.

Some head-cleaning cassettes (particularly the "dry" kind) work by abrasive action, so naturally you'll want to take it easy and avoid using these tapes until you really need them. It's possible to wear down a video head by using a cleaning

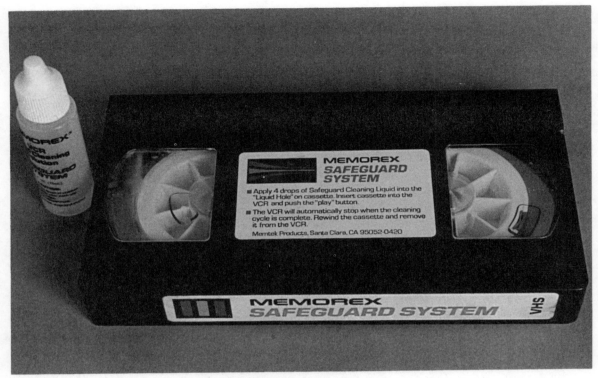

Fig. 5-21. A commercially available wet/dry head-cleaning system.

cassette, so follow directions carefully and write down the date whenever you use the tape. If you notice that you're cleaning the heads more than every one month or so, you might have other problems, like an old tape that's shedding too much of its oxide coating. More about this is later in the chapter.

A growing number of cleaning cassettes use the "wet" or "wet/dry" method. Before using the cassette, you apply a drop of a liquid cleaner to the tape. When you insert the cassette, the wet tape cleans the heads and any other metal or plastic part it touches. The problems of overuse and abrasion are not as great with this type of cleaning tape, but use it carefully nonetheless.

Normally, head-cleaning cassettes do a fair job at cleaning the spinning video heads as well as the stationary heads and other metal parts inside the VCR. A few brands do not clean all the stationary heads, however, nor the entire tape path.

True, the stationary heads do not need as stringent a cleaning as the video heads, but an excessive build-up of dirt and oxide can lead to sound distortion, inadequate erasing, and faulty picture synchronization. If your favorite cassette doesn't clean the entire innards (as described in its instruction manual), consider supplementing it with another brand or method.

Manual Cleaning

The other video-head cleaning method involves manually cleaning th
You need two supplies: cleaner and cleaning pads. A Freon TF and alcohol
(available at any electronics store) or other recommended video-head clea
solution is ideal for cleaning video heads. Isopropyl alcohol can also be used,
long as it is "technical grade" or "medical grade"—97 percent alcohol or more.
The higher the percentage of alcohol, the less water is in the mix. The water takes
longer to evaporate, making maintenance more difficult, and can prevent adequate
removal of oxide and other contaminants. Don't use a cleaning solution designed
for audio heads. These sometimes have lubricants that aren't needed or desired
for video heads.

Be especially wary of head cleaners that you spray on, though these can be
used. The cleaning agent in these is Freon, which is very cold (it actually reaches
its boiling point when it hits the air). The sudden blast of cold can easily crack a warm
or hot video head, rendering it useless. If you use a spray-on head cleaner, be sure
the VCR has cooled before you spray.

Apply the cleaner with a chamois-tipped swab or pad only—*never use cotton
swabs as the lint can come off and become entangled in the head mechanism.* You

Fig. 5-22. Cleaning the video heads with a chamois-tipped applicator. Remember to fully wet
the chamois with the cleaner.

129

Video Heads

virgin chamois, cut into a convenient square-inch
ily wet the swab and *gently* rub it sideways against
Never use an up-and-down motion.—you will sure-
off. Most VCRs have several video heads, so be

eads is to apply the swab (or sponge or chamois)
rotate the top portion of the drum manually with
y. 5-23). Apply only moderate pressure on the cleaning pad. When
using a chamois pad you will feel the head pass under your finger.

After cleaning, wait at least five minutes before inserting a tape and playing it.
The waiting period is important; it allows the cleaner to fully evaporate. If you play
a tape and the heads are still wet, they will get dirty again, and the oxide residue
will be even harder to remove.

Throw the cleaning swab or chamois away after use. Although it may still seem
clean, re-using it can impair the effectiveness of future cleanings.

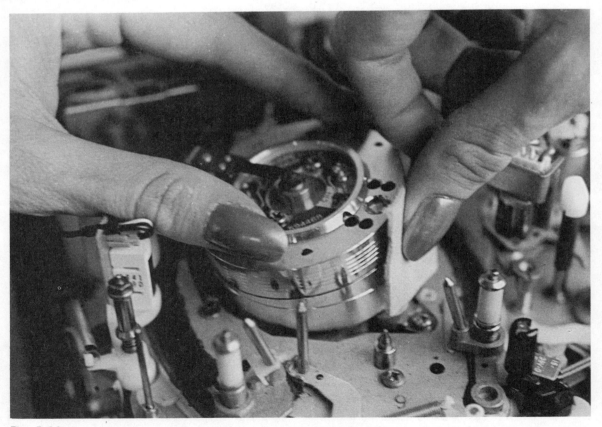

Fig. 5-23. Press a small square of chamois against the head drum. Move the heads back and forth by rotating the
top part of the drum with the other hand.

Sometimes, the heads can be so dirty that ordinary cleaning won't remove the oxide build-up. If the picture won't clear after two or three cleanings (wait the recommended time between each cleaning), try a dry-type cleaning tape. The "lapping" action of the tape will likely scrub off the oxide and clear up the video.

Note that all VCRs (except for Betamovie camcorders) use at least two heads. Depending on the model, format, and whether it has hi-fi stereo, there can be as many as seven heads on the drum. You can sometimes determine the function of each head by reading the labeling on the top of the head drum; otherwise, consult the service manual for your machine. In any case, be sure that you get all the heads. The entire process might take five or ten minutes if your deck is equipped with four or more heads.

Cleaning Other Heads

While you're inside the machine cleaning the video heads, get all the other heads as well. This includes the full erase head and the main head unit, which includes the contract track, audio, and (possibly) the audio erase heads. Be sure to thoroughly clean each metal part that comes into contact with the tape, including guide rollers, impedance rollers, treading guides, even the entire video head drum surface (this removes gunk that can cause friction on the tape). You can use the cleaner on hard plastic parts (oxide build-up is especially noticeable on these), but avoid applying the cleaner on soft rubber parts. These require a special rubber cleaner, available at most electronics stores.

How Often Should You Clean the Heads?

Some experts say you should clean the heads only when the playback picture shows a definite loss in clarity or shows an increase in snow. Others say to clean every 75 to 100 hours of use. Still others claim you should do it every 25 to 35 hours of use. Who's right? They all are. It's what you're comfortable with, what you need, and what the manufacturer recommends for its product.

The thing to remember is that the heads usually do not require cleaning every day or even every week. With a VCR that receives moderate use, cleaning once every month or two is reasonable.

Cleaning Costs

Prices for cleaning kits—either cassettes or swabs—vary. A pack of four to six swabs and a bottle of cleaning solution costs from $6 to $10. Cleaning cassettes range from $12 to $30. Either kit is the best investment you can make to keep your VCR alive and well.

OILING AND LUBRICATING

Video cassette recorders generally require little lubrication. After a year or two of use, however, it is advisable to apply small amounts of oil and grease to keep

GENERAL CLEANING AND PREVENTATIVE MAINTENANCE

the deck in top-notch condition. At every major PM interval, you should lubricate such parts as the shaft of the capstan pinch-roller, the threading mechanism, the loading mechanism, and the mechanical controls. Lubricate these parts more often if you use your deck outdoors where it is subject to extremes in temperature or excess dust and dirt.

You should also lubricate the mechanical parts if you clean the machine with a spray cleaner/degreaser. The cleaner acts to strip any oil or grease from the moving parts, and using the VCR without lubrication can wear it out prematurely.

Here's a good rule of thumb to remember when lubricating your VCR:

◆ If it spins, oil it;
◆ If it slides or meshes, grease it.

Light machine oil is the best lubricant for spinning parts; such as the bearings of idlers and rollers. You should *never* use a spray-on lubricant, such as WD-40. Go easy on the oil and apply it only to the rotating part. Never apply oil to the surface of a roller or where the lubricant might come into contact with the tape. Remember, more is not necessarily better. Excessive oil will gum up the tape causing clogged video heads, jammed tapes, and other maladies. Never apply oil directly to a motor. For small parts, apply the oil with a syringe applicator, as shown in Fig. 5-24.

Fig. 5-24. Apply a small drop of oil with a syringe or needle applicator to only those parts that absolutely require it.

132

Fig. 5-25. Grease should be applied to parts that slide or mesh, such as the tape spindle slides in a VHS deck.

A non-petroleum-based grease is the best lubricant for sliding and meshing parts, the bulk of which are found in the loading and threading assemblies, and in mechanical controls (older decks only). This includes cams, bell cranks, gear surfaces, and levers. As with oiling, apply only a small dab of grease. Spread the grease by moving or sliding the part back and forth a few times.

If the part can't be easily moved or is geared down and won't budge, power the deck and insert a tape. To spread the grease, cycle the loading mechanism by inserting a cassette and using the front panel controls of the deck.

Use grease to lubricate the slides in the tape threading mechanism. Apply a dab of grease to the end of the slide (see Fig. 5-25) and cycle the deck between play and stop modes to evenly spread the lubricant (with Beta decks, repeatedly eject and re-insert the tape to activate the threading mechanism). You will need to insert a cassette into the deck or the VCR won't respond. Use a cotton swab to soak up excess grease. If the slide is dirty, clean it first before adding more grease.

Grease should also be applied to the mechanical controls on older VCRs (these machines have ''piano-key'' controls). Locate the levers and mechanisms for the PLAY, FAST-FORWARD, REWIND, and other controls and apply a small amount of grease. Press the controls to spread the grease and wipe up the excess.

GENERAL CLEANING AND PREVENTATIVE MAINTENANCE

Grease and oil should also be applied to the gears on the underside of the threading mechanism. In most VHS machines, the threading mechanism is driven by worm gears. These require grease to keep them from binding. As before, apply a small dab of grease and cycle the threading mechanism back and forth to distribute the lubricant.

Another sub-system of the VCR that needs lubrication is the cassette-lift mechanism. This applies to both top- and front-load VCRs. Apply a small dab of oil to the shafts of the gears and pulleys. Avoid using too much oil; it might find its way to the tape and cause problems later.

TUNER CLEANING

Mechanical detent tuners on old-model VCRs need regular cleaning (about every 24 to 48 months). If the tuner isn't excessively dirty, you can apply spray cleaner to it without removing any of the mechanism. Use a tuner cleaner/lubricant (available at Radio Shack) and squirt it liberally inside the tuner mechanism. Spin the tuning dials quickly in both directions to spread the cleaner.

If the tuner is very dirty, you might have to remove it from the machine to gain adequate access. Remove the front-panel knobs and locate the screws that hold the tuner in place. Remove them and gently lift the tuner out. The tuner will be connected to the VCR by several wires; if the deck uses connectors, unplug the wires and note where they belong. If the wires are permanently soldered in place, remove the tuner only enough to gain access to the internal contacts. Liberally spray the cleaner/lubricant, then replace the tuning mechanism.

Note that in the older VCRs, the tuner mechanism is composed of many channel strips—one strip for each VHF channel. Each strip tunes a specific channel. If the tuner is out of adjustment, the strips might need to be serviced or recalibrated. Although this requires some special tools and test tapes, it can be done at home if you follow the instructions in the service manual.

DEMAGNETIZING

Over time, the heads and other metal parts that come in contact with video tape will build up a small amount of residual magnetism. This magnetization can impair the signal quality of your tapes, resulting in a muddy or garbled sound and sometimes a weak picture. To return good sound to your tapes, regularly demagnetize the audio heads, the tape path, and all ferrous metal parts that come in contact with the tape.

Video heads can also be demagnetized, although most VCR manufacturers now recommend against it, saying it is unnecessary. You can do it as long as you use a demagnetizing tool specially made for use in VCRs and on video heads. Use of a regular reel-to-reel tape recorder demagnetizer can shatter the video heads, due to the intense magnetic field that is induced by the tool. If you decide to demagnetize the video heads in your VCR, proceed at your own risk.

To use the demagnetizer, start with the tip about three feet from the deck. Turn on the demagnetizer, and slowly bring it close to the VCR. Wave the tip past the full erase head, tape guide spindle, audio heads, video heads, and other components. [*Do not physically touch* any of these parts, or you can leave more residual magnetism than you left.] Be especially wary of touching the video heads with the tip of the demagnetizer. Even with the proper tool, actually touching the heads can shatter them.

When you are done, keep the demagnetizer on and slowly draw the tool away, again to about a three feet distance. Turn the tool off. If power to the tool is turned on or off when the tip is closer than one or two feet from the deck, repeat the process. Otherwise, you will leave the deck magnetized.

Be aware that the demagnetizing tool can erase your cassettes. When using the tool, place all tapes away from the work bench.

END-OF-TAPE SENSOR CLEANING

The end-of-tape sensor detects the clear leader portion spliced onto the beginning and end of all VHS and 8mm video tapes. An infrared (sometimes visible) light shines through the leader and strikes a photodetector, as shown in Fig. 5-26. When the photodetector sees light, it instructs the deck to stop playing (it also works during rewind and fast-forward).

If the sensor is dirty, it may never detect the light, and you loose the auto-stop safety feature. This can lead to stretched or broken tapes. Use a cotton swab dipped in alcohol to clean the sensors and the lamp assembly. In most all VCRs, there are two sensors, located on either side of the loading/threading mechanism. Be sure to clean them both.

Beta VCRs use electro-magnetic coils to sense the ends of the tape. Gently clean the coils with a brush and inspect the bracket that connects them to the deck.

REEL SPINDLES

The reel spindles rotate to supply and take up tape as it travels through the tape transport mechanism. Although you wouldn't know it, the action of the spindles is

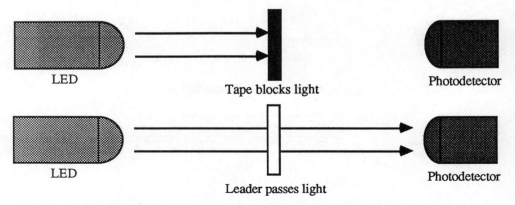

Fig. 5-26. The operation of the end-of-tape sensor in a VHS recorder.

GENERAL CLEANING AND PREVENTATIVE MAINTENANCE

very delicate, and excessive dust or dirt can impair their performance, leading to considerable problems.

Use a soft brush to get rid of an accumulation of dust and grime. Apply a small amount of oil to the top of the spindle. Use only a small amount. Rotate the spindle to spread the oil on the shaft. Only if the spindle is exceptionally dirty or jammed should you remove it. The height of the spindle is precise and on most machines is determined by the number and thickness of small washers installed on the shaft. If you do disassemble the spindles, be sure to put them back together in the exact same sequence.

Note that under the take-up spindle there's a sensor. This sensor uses light to detect when the spindle is turning. If the spindle stops turning during playback (an event that can lead to tape spillage), the sensor relays this information to the VCR and the VCR shuts itself off after 5 or 10 seconds. Because this sensor uses light, any dust, oil, or grime can impair its performance. While you have the take-up spindle off, use a brush to clean the sensor and the inside rim of the spindle.

FRONT PANEL CONTROLS

The front panel operating controls seldom need preventative care unless the VCR has been subjected to heavy doses of dust, dirt, damp air, and cigarette smoke. You can clean the bulk of the control switches with a brush or compressed air. The control panels in most decks use sealed membrane switches, so the internal switch contacts are not effected by dirt (the membrane switches are located behind the actual switch button, as illustrated in Fig. 5-27, which you push when operating the player).

If the switches in your VCR are the regular open contact type, you might need to apply a small amount of electrical contact cleaner to dispel the dirt. Again, you need not perform this procedure unless the deck has been exposed to heavy amounts of contaminants or the controls are not working properly.

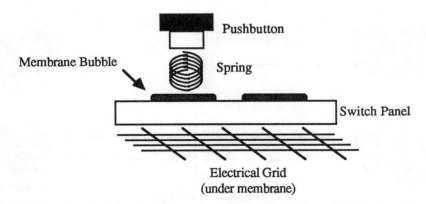

Fig. 5-27. Membrane switches don't use exposed electrical contacts, so cleaning is not required. The contacts of the switches are located under or within the membrane bubble.

The contact cleaner dries quickly and without leaving a residue. Nevertheless, you should wait several minutes after cleaning before you turn the machine on and attempt to use it. As usual, never clean the switch contacts when the machine is turned on or plugged into the wall socket.

FINAL INSPECTION

Before you replace the top and bottom covers, double-check your work. Make absolutely certain you haven't left tools, hardware, used cotton swabs, or cleaning supplies inside the chassis of the VCR. Check to make sure that you have properly replaced *all* parts.

REASSEMBLY

If everything looks satisfactory, replace the covers and reassemble the deck with the screws you removed. Be sure the covers are aligned properly before tightening any of the screws. When all the screws are in place, tighten them lightly. Do not overtighten or you might strip the threads of the screw and chassis.

CHECKOUT

After every PM interval, check the deck for proper operation. You should use one known good tape (either a commercial tape or one you've made—see Chapter 4 for details). Keep these operational points in mind during the checkout process:

- ✦ The tape-loading mechanism operates smoothly and accepts the tape as it should.
- ✦ The tape threads and unthreads without mishap (try the sequence several times).
- ✦ The tape plays properly, with full audio and video.
- ✦ The FAST-FORWARD and REWIND buttons operate properly and shuttle the tape to the end.
- ✦ The deck automatically shuts off when it reaches the end of the tape (and the same function should work when rewinding the tape).
- ✦ Operating the tuner selects the different channels.
- ✦ All front panel indicators operate normally.
- ✦ All the function buttons on the front panel of the deck and the remote control work as they should.

MAINTENANCE LOG

You might want to keep a logbook of all the maintenance checks and cleaning you've performed on your VCR. Use the sample log form that appears in Appendix C, or make your own. In your log, be sure to note the exact checks and maintenance procedures you performed and when you performed them. If you have replaced any parts (like tires and belts), note their part numbers in the log and troubles you

GENERAL CLEANING AND PREVENTATIVE MAINTENANCE

might have had when installing them. By keeping a thorough log, you can better estimate the timing of the preventive maintenance intervals for your specific deck.

REMOTE CLEANING

Almost all VCRs are equipped with wireless infrared remote control units. These allow you to operate the deck from afar—play a tape, scan through a tape, and more—all from the comfort of your easy chair. Because the remote control can be taken anywhere in the house, it's often subjected to a lot more abuse than the other components in your video system. You might pride yourself in a clean VCR, but the remote could be filled with dirt, grime, or sticky syrup from a spilled soda pop.

If your VCR has a wired remote, you should follow the same general cleaning procedures. Also note that you should inspect and repair any damage that might have occurred to the wire leading between the remote and VCR.

Exterior Cleaning

Cleaning the remote control is straightforward and easy. Start with applying a damp cloth to the outside and remove exterior dirt and grime.

Battery Compartment Cleaning

Remote controllers are battery operated. Given the right circumstances, all batteries can leak, and the spilled acid can corrode the electrical contacts and can impair proper operation of the controller. If your remote control is operating sporadically, this might be the problem.

If the batteries have leaked or if the battery contacts have become only slightly dirty, clean them as follows. You can also use this procedure for cleaning the battery compartment in portable VCRs and camcorders.

◆ Carefully remove the batteries and discard them. Immediately wash your hands to remove any acid residue. Battery acid burns.
◆ Wipe off the battery acid residue with a damp cloth, and remove as much of the excess from the contacts as you can.
◆ As shown in Fig. 5-28, remove the remaining residue by rubbing the contacts with the tip of a pencil eraser. Blow the eraser dust out of the battery compartment when you are through.

Test the remote with new batteries. If the battery leakage is more extensive, you will have to take more drastic action. Refer to Chapter 6 for more details on how to clean heavy deposits of battery acid.

Interior Cleaning

To clean the inside of the remote, you must disassemble it by removing the screws holding the two halves together. The screws are small and you will need a

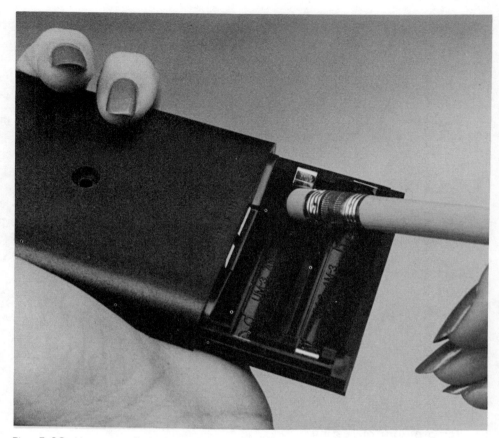

Fig. 5-28. Use a pencil eraser to clean the battery contacts.

set of jeweler's screwdrivers to properly remove them. Some screws may be hidden in the battery compartment, under the manufacturers label, or even beneath the thin metal cover on the front of the unit.

When the screws have been removed, separate the two halves. Clean the inside of the controller with a soft brush or can of compressed air. If liquids have been spilled inside the controller, spray the inside of the controller with a suitable cleaner/degreaser.

With most remote controllers, the circuit board is attached to the front panel and must be removed if you want to access the front switches. With the remote resting on a flat surface, remove the circuit board screws carefully and gently lift off the control board. The plastic pushbuttons might be loose, so be sure not to knock them out of place when you remove the board (if so, putting them back isn't hard, but it can be time consuming). Clean the switches and front side of the circuit board with a spray from the can of cleaner/degreaser, as shown in Fig. 5-29.

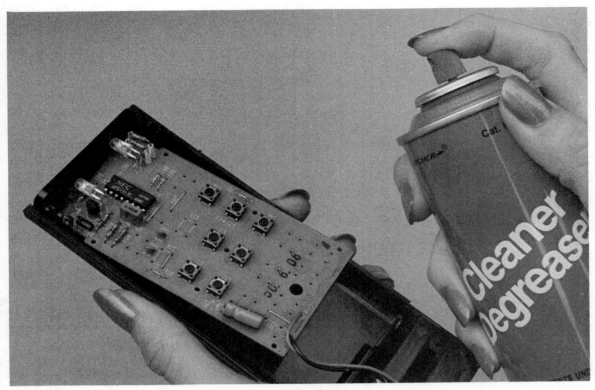

Fig. 5-29. Spray cleaner can be applied to the switches in the remote control.

Remote Control Testing

Let's say you've replaced the battery in the remote controller and you've cleaned it inside and out. It still doesn't seem to work. Since the infrared light from the remote is invisible, it's difficult to test that it is indeed working or if the problem lies in the VCR itself.

Many remote controllers have visible LED indicators that flash when you push a button on the transmitter. The LED is typically connected with the high-output infrared LEDs that send the remote control information to the player. You can be fairly certain that the controller is operating properly if the visible LED flashes.

Still, like everything, the infrared LEDs in your remote controller do not last forever. It's possible, but not really likely, that one or all of the high-output infrared LEDs have gone out. If only one of the several LEDs that are typically built into the controller is faulty, you should still be able to operate the remote, but you might have to get in close to the deck.

One way to accurately test the remote is to use the infrared sensor described in Chapter 4. This sensor is sensitive to infrared light and will register the light emitted by the remote controller. To use the sensor, position it closely to the front of the controller and start pushing buttons. If the LED on the sensor lights up, the remote is good. The fault probably lies in the deck. If it does not light up, double-check the operation of the sensor as described in the last chapter and try again. If the LED on the sensor still doesn't light up, you can be fairly certain the remote control is at fault.

Lacking the infrared sensor, you can use an ordinary AM transistor radio to test the remote. Dial the radio to 530 on the dial (or thereabouts if there is a station broadcasting on that frequency) and place it near the front of the remote. Push the buttons—you should hear a chirping sound if the remote is working. Note that this test does not actually verify light output from the remote but that the electronics in the remote are working.

Be aware that there are a variety of reasons why a remote control does not operate properly, and none of them have anything to do with a faulty controller or sensor in the VCR. If you are having trouble with your remote control and have performed the routine cleaning outlined here, check the remote control troubleshooting charts in Chapter 9 for more information.

TAPE TROUBLES

Sometimes sound and picture ailments are not the fault of the VCR, but the videotape. Here's how to make sure:

Old Tapes

Be on the lookout for troubles arising from aging, worn-out tapes. Older tapes have a tendency to shed their magnetic oxide coating. This can cause dropout (the small white flecks that dash randomly across the screen). Circuitry inside the VCR compensates for some of the dropout, but a badly deteriorated tape can be unwatchable.

If one of your tapes gets so bad that it's difficult to view, these steps should be followed.

- ➡ Make sure it's not the VCR's fault by testing the machine with a known good tape.
- ➡ If it is indeed the tape that's bad, as a second step, cover the cassette with a white sheet, say a few nice words, and chuck it.

Stretched Tapes

Old or mishandled tapes can stretch, which causes the tape to curl. This in turn creates several objectional problems, one of which is a bending or hooking of the picture at the top of the screen. Badly stretched and curled tapes can bring about other malfunctions, such as poor audio and a loss of picture synchronization. Unfortunately, like old and shedding tapes, little can be done to help "unstretch" a tape.

Jammed Tapes

VCRs that have not been maintained have been known to ''eat'' tapes. During recording or playback or during threading, the tape jams within a VCR. If a jam occurs, immediately stop the VCR and press the EJECT button.

The tape might be tangled; if it is, turn off the deck and remove the top cover. Manually extract the tape, being careful not to touch it. If you have a pair of clean cotton gloves, now is the time to wear them. If you don't have gloves, handle the tape by the edges only, because oil from your skin can cause head clogging and further tape damage.

When the tape is clear from all obstructions, take the cassette out of the loading port, if you haven't done so already. If the cassette cover closes on the tape, you'll need to reopen it by inserting a flat-bladed screwdriver into the cover release button, as shown in Fig. 5-30.

Both VHS and Beta tapes have locks to prevent the reel tables from turning when the tape is removed from the machine. Opening the door on the Beta tape will automatically disengage the reel locks. VHS tapes have a small button on the bottom (Fig. 5-31) that must be depressed to release the locks (8mm tapes have a similar button on the bottom). Do not attempt to force the reels to turn—you'll snap the tape in two.

If the tape is severely damaged, it might be permanently ruined. You can cut out the bad portion and splice the tape back together, but this should be done at the beginning of the tape only. See the next section for more information on tape splicing.

Splices

Unlike audio tape, it is not advisable to splice videotape except for the clear leader portion at the beginning. Splices along the length of the tape aren't recommended because they can cause damage to the spinning video heads. Since the leader on videotape never passes by the heads, it is permissible to splice the tape in this area.

To splice videotape, gather the necessary tools around you: a pair of cotton gloves, a splicing block for ½-inch tape (or 8mm tape if you are splicing an 8mm cassette), a razor blade, and splicing tape (DO NOT use cellophane tape).

➡ Pull the tape out of the cassette until all of the damaged parts are exposed. Keep the cassette door open with a piece of masking tape or rubber band. Be sure to unlock the reels when you are pulling the tape out or you might stretch it.

➡ Cut out the bad portions and throw the tape away.

➡ Lay the two ends of the tape on the splicing block as shown in Fig. 5-32. Lay the tape so that the shiny side is up (if both sides appear shiny, orient the tape so that the surface on the inside of the wound reels is facing you).

➡ Overlap the tape slightly and cut both strands with the razor blade. Most

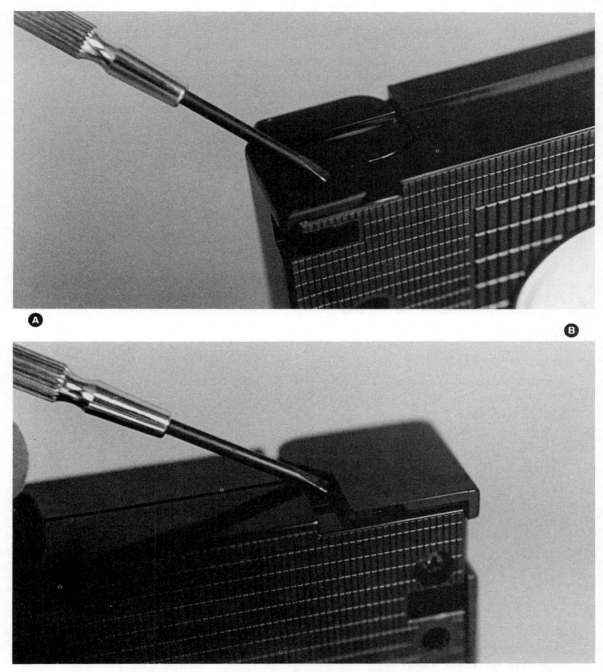

Fig. 5-30. A) The location of the cover release pin on Beta cassettes. B) The location of the cover release pin on VHS cassettes.

Reel Unlock
Button

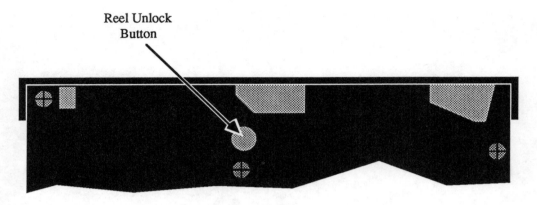

VHS cassette (bottom view)

Fig. 5-31. The reel unlock button on VHS cassettes is located on the bottom, inside a large hole.

splicing blocks have several cutting grooves: for splices in the middle of the tape use the diagonal groove; for end cuts use the diagonal one.

➡ Immediately splice the two ends together with tape. Be sure that the tape ends don't move. Avoid gaps or overlap in the splice, as shown in Fig. 5-33.

Tape Shell Damage

The cassette cartridge itself (called the shell) might become damaged for a number of reasons: it could get crushed when stored in the bookshelf, broken in

Splice Grooves

Splice Block

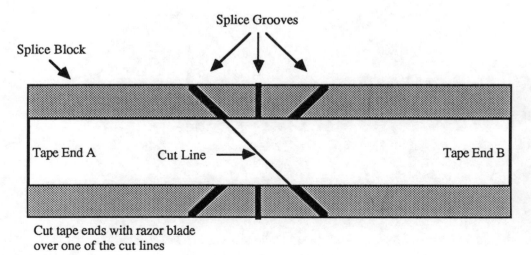

Tape End A   Cut Line ——→   Tape End B

Cut tape ends with razor blade
over one of the cut lines

Fig. 5-32. Use one of the diagonal grooves in the splicing block to trim the tape ends.

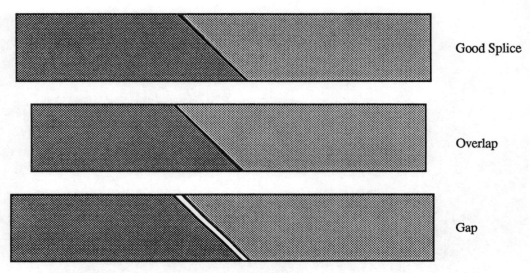

Good Splice

Overlap

Gap

Fig. 5-33. Good splices are devoid of gaps and overlaps.

transit through the mail system, snapped apart when packed in the suitcase of your latest worldwide tour, or warped by the heat of the sun. You can sometimes repair a cracked shell by using modeller's cement. If the shell is beyond repair but you want to save the tape that's inside, you can transfer the tape from one cassette body to another.

Using Modeller's Cement. Small cracks can be fixed by applying liquid modeller's cement to the shell. Use a small "00" or smaller brush to apply the cement. Dip the brush in the bottle and lightly touch the tip of the brush against the crack. Capillary action will draw the cement into the crack and spread it. Do not apply too much cement or it might drip onto the tape inside. You can also use Super Glue or other strong adhesive, as shown in Fig. 5-34.

Transferring Tape to Another Shell. To replace the shell, you need an extra tape that you don't care about losing. A bargain basement tape recently bought at the five-and-dime store or a worn-out tape can be used. All you are interested in is the shell, so the condition of the tape itself is of no consequence.

Use a *de*magnetized screwdriver to loosen the screws on both cassettes. Open the shells on a flat surface, like a large table. Be sure that none of the internal parts, like rollers and springs, fall out. If they do, quickly note their position and replace them. When disassembling an 8mm cassette, use a jeweler's screwdriver.

Examine the way the tape threads around the rollers, posts, and other parts of the shell. Make a diagram if you think you might forget. Figure 5-35 shows a VHS tape opened up, revealing the threading course.

Take the two reels of tape from the good shell and gently lift them out. Place the tape aside or throw it away. Now, lift the two good reels out of the bad shell and place them in the good shell. Carefully thread the tape around the posts and rollers, and reassemble the shell. When replacing the screws, do not over-tighten.

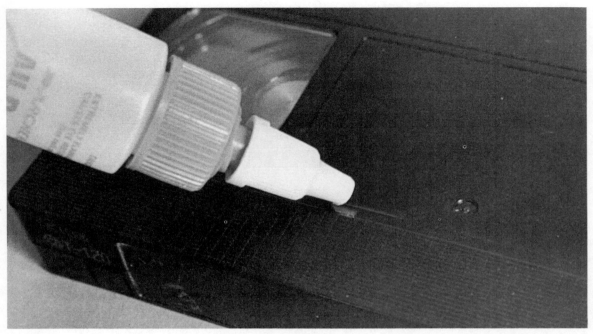

Fig. 5-34. Apply a small amount of solvent-based cement to repair cracks and breaks in plastic pieces and tape cassettes.

Fig. 5-35. A VHS cassette, opened to reveal the supply and take-up reels.

With most shells, the screws are self-tapping and you'll strip out the plastic if you exert too much pressure.

Test the transplant by playing the tape on your VCR. Listen for squeaks or odd rubbing sounds—tell-tale indicators that the tape is not properly threaded within the shell. If so, take the shell apart and try again.

Tape Care Guide

Follow these simple guidelines to maximize the health and longevity of your video tapes.

- ✦ Store all tapes in an upright position (see Chapter 3 for more details.)
- ✦ Store tapes in a cool, dry place, away from direct sunlight.
- ✦ Avoid placing tapes on floor or carpet where they might pick up dust and fibers.
- ✦ Never touch the tape surface; wear gloves when handling or splicing tapes.
- ✦ Place tapes several feet from electro-magnetic sources. These include loudspeakers, TV sets, and telephones.

Light Use

PM (Months)	3	6	9	12	15	18	21	24	27	30	33	36
Clean/Dust Exterior/Interior	□	□	■	□	□	■	□	□	■	□	□	■
Clean Heads	□	■	□	■	□	■	□	■	□	■	□	■
Lubricate Transport	□	□	□	□	□	■	□	□	□	□	□	□
Clean Rubber Parts	□	□	□	□	□	□	□	□	□	□	□	□
Clean Remote Battery Terminals	□	□	□	■	□	□	□	■	□	□	□	■
Clean Connectors	□	□	■	□	□	■	□	□	■	□	□	■

Medium Use

PM (Months)	3	6	9	12	15	18	21	24	27	30	33	36
Clean/Dust Exterior/Interior	□	■	□	■	□	■	□	■	□	■	□	■
Clean Heads	□	■	□	■	□	■	□	■	□	■	□	■
Lubricate Transport	□	□	□	■	□	□	□	■	□	□	□	■
Clean Rubber Parts	□	□	□	■	□	□	□	□	□	□	□	□
Clean Remote Battery Terminals	□	□	■	□	□	□	□	■	□	□	□	■
Clean Connectors	□	■	□	■	□	■	□	■	□	■	□	■

Heavy Use

PM (Months)	3	6	9	12	15	18	21	24	27	30	33	36
Clean/Dust Exterior/Interior	□	■	□	■	□	■	□	■	□	■	□	■
Clean Heads	■	■	■	■	■	■	■	■	■	■	■	■
Lubricate Transport	□	■	□	■	□	■	□	■	□	■	□	■
Clean Rubber Parts	□	■	□	■	□	■	□	■	□	■	□	■
Clean Remote Battery Terminals	□	□	■	□	□	□	□	■	■	□	□	■
Clean Connectors	■	■	■	■	■	■	■	■	■	■	■	■

Your Machine

PM (Months)	3	6	9	12	15	18	21	24	27	30	33	36
Clean/Dust Exterior/Interior	□	□	□	□	□	□	□	□	□	□	□	□
Clean Heads	□	□	□	□	□	□	□	□	□	□	□	□
Lubricate Transport	□	□	□	□	□	□	□	□	□	□	□	□
Clean Rubber Parts	□	□	□	□	□	□	□	□	□	□	□	□
Clean Remote Battery Terminals	□	□	□	□	□	□	□	□	□	□	□	□
Clean Connectors	□	□	□	□	□	□	□	□	□	□	□	□

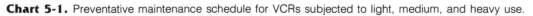

Chart 5-1. Preventative maintenance schedule for VCRs subjected to light, medium, and heavy use.

6

VCR First Aid

Accidents do happen to even the most careful people and to their VCRs. If an accident befalls your video cassette recorder—you spill your cup of coffee into it, for example—there is no need to panic. By following some simple cleaning and checkout procedures, you can reduce or eliminate potentially costly repairs.

This chapter discusses first aid treatment for a number of accidents and how you can repair all or nearly all of the damage yourself. Even if your VCR is presently healthy and working fine, read through this chapter to acquaint yourself with the recommended procedures. You never know when they'll come in handy.

TOOLS AND SAFETY

Most of the repair procedures require disassembly of the deck. If you haven't done so already, please read Chapter 4 and 5 for more information on how to take your VCR apart, and what tools you need to do the job right. Pay particular attention to the safety points in Chapter 5.

Bear in mind that the steps outlined in this chapter are designed to help you minimize the damage of accidents, not make them worse. If you are not confident in your ability to perform some of the repair procedures or lack the proper tools and cleaning supplies, then by all means don't do them. When in doubt, refer to a qualified technician.

DROPPED DECK

VCRs and camcorders are not meant to be dropped off bookshelves, TV racks, tables, or car seats, but it happens. A drop onto soft carpeting might just shake up the deck a bit but not cause serious internal injury. A harder impact, however, can break or bend the cabinet or chip off a piece of the front panel.

A cracked cabinet or other broken piece on the exterior of the machine can usually be mended with a strong adhesive. If the part is plastic, just about any clear glue recommended for use on plastic will suffice. After gluing, hold the pieces together with your fingers until the adhesive sets. Or, tape the pieces together like a surgical suture until the glue is completely dried.

Checking for Internal Damage

After a VCR has been dropped, even if there is no visible damage, inspect it carefully. Don't test it by plugging it in and playing a tape. You might cause additional damage.

You can quickly check if parts have come loose inside the deck by unplugging the unit from the wall socket and gently tilting or shaking it. If you hear rattles, you can bet something is bouncing around that shouldn't. Even if you don't hear loose parts, it's a good idea to disassemble the VCR before attempting to use it. Follow the general disassembly instructions provided in Chapter 5.

When the top cover of the cabinet is removed, look for the loose parts and ascertain where they came from. If the broken piece is plastic and is not an extremely crucial part, glue it back on with a suitable cement. A solvent-based cement can be found at hobby or plastics supply stores. Broken metal parts are harder to mend, but there are a number of adhesives for metal that might do the job. The alignment of internal parts is often critical to the operation of the VCR, so make sure you glue the repaired piece on straight.

While the top is off, visually inspect the interior of the deck for hidden damage—things other than completely broken pieces. Pay particular attention to circuit board(s), the tape loading mechanism, the tape threading mechanism, and the video head drum. If any part of the works looks broken, bent, or out of place, you should return the deck to a qualified repair center. A broken head cannot be serviced; it must be replaced, and replacement requires exact alignment using special tools.

Check the printed circuit boards inside the VCR. If you see any hairline cracks, it's a bad sign. A broken printed circuit board must be replaced. Using the deck as it is can cause more damage, because components might be shorted. You could conceivably burn out the power supply, motors, and other costly parts by attempting to operate the VCR with a faulty circuit board. Fortunately, however, it takes a very healthy jolt to break a circuit board, so it is not a common occurrence.

Broken Wires

A strong enough impact can cause electrical wires and connectors to break or come off. Sometimes the connector, like the one in Fig. 6-1, is just jostled in it's socket. Look very carefully at these, because some connectors might look fine from the outside but electrical contact inside has been broken. Press all the connectors firmly

Fig. 6-1. A ribbon cable and connector attached to a circuit board.

into their sockets just in case. If the connector has come out all they way, just plug it back in to its respective socket.

Broken wires leading to connectors and boards must be resoldered. If the solder joint is accessible and does not require precision work, you can do the work yourself. If you are unfamiliar with proper soldering techniques, refer to Appendix D for a quick lesson.

If more than two wires have come undone and it is not obvious where they go, consult the schematic diagram for your VCR, if available, or return the deck to a service shop. Don't take chances here. If two wires are loose, you have a 50 percent chance of soldering them back properly. You also have a 50 percent chance of soldering them back in the wrong place—not particularly great odds, considering the permanent damage that can ensue.

Leakage Current Test

Prior to turning on your VCR, even if you do not take it apart to inspect for internal damage, be sure to check for possible leakage current. This test determines if any part of the ac line has come in contact with the metal cabinet or base. It is a safety check to prevent a potential shock hazard. You'll need a volt-ohmmeter to perform the test.

Method #1:

With the VCR unplugged, short the two flat prongs on the end of the ac cord with a jumper wire, as illustrated in Fig. 6-2. On the meter, select the ohm function and a range of no less than 1 KΩ ohms. Connect one test lead from the meter to the jumper on the ac cord. Connect the other test lead to any and all bare (not painted) metal parts of the deck. For a typical VCR, the meter should read about 50 KΩ to 100 KΩ—if you get any reading at all.

A reading substantially lower than this is a good indication that the power supply has come into contact with the metal parts of the the deck. If this happens, inspect the power cord as it leads into the deck, as well as the ON/OFF switch, the transformer (usually bolted onto the back of the machine), and the wires leading from the transformer to the printed circuit board. If the wires are broken or are shorted to the cabinet, repair the fault before using the deck.

VCR (Back View)

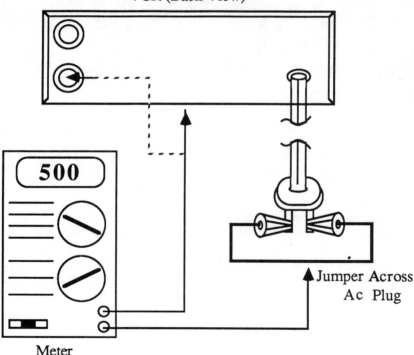

Jumper Across
Ac Plug

Meter

Fig. 6-2. One method for testing leakage current with a volt-ohmmeter.

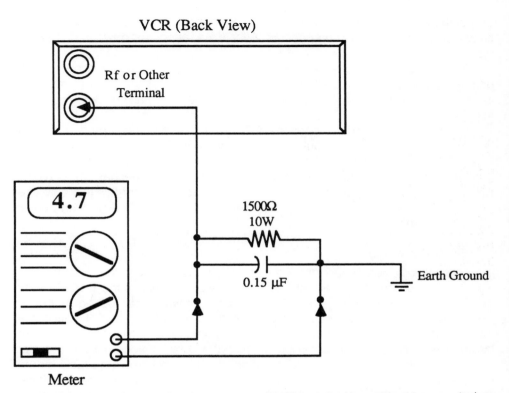

Fig. 6-3. Alternate method for testing leakage current. This method is preferred because it gives an actual reading of the current.

Note that not all exposed metal parts of the VCR will return a high value. Touching the center conductor of one of the audio output connectors could yield a very low resistance of 1 KΩ to 50 KΩ. This is normal with some machines.

Method #2:

Follow the circuit diagram shown in Fig. 6-3. Place the meter in the ac millivolts range and take a reading. The value should be very low—a matter of 2 to 10 millivolts. Anything higher might indicate a partial or complete short circuit. Note that you need to connect one side of the meter to a suitable earth ground. The ground connector in a wall outlet is a good choice. If your cable TV system is adequately grounded (the outer conductor of the coaxial cable is attached to a grounding rod), you can use it instead.

FIRE DAMAGE

Fire damage includes the effects of both the heat and the smoke of fire. Intense heat can totally destroy the VCR, of course, but even moderate heat (150 to 175 degrees) for even a short period time can cause considerable damage to the deck's electrical and mechanical components. After a fire, check for obvious damage to plastic parts.

If a tape was in the deck, remove it and inspect it for warpage, melting, or other damage. If it is warped, then there is a good indication that the other plastic parts inside the deck are warped as well. Disassemble the machine and inspect it thoroughly.

Minor warpage of plastic parts, as well as expansion of metal parts during the heat of a fire can throw components out of alignment. You can test this by playing a tape from beginning to end (you can fast-forward through the tape to save time). If anything is out of alignment, such as the tape threading mechanism, playback will be impaired for at least some portions of the tape. Perform this test only after thoroughly inspecting the deck and are certain that powering it up won't cause additional damage.

Smoke Damage

Many people erroneously think that heat from a fire causes the greatest damage. Insurance companies will tell you that smoke, not heat, is the worse enemy. Why? The smoke gets everywhere, even rooms that were not touched by the fire itself. The black soot of smoke cakes on everything and has an acid-like effect that can eat through finishes and protective layers. If your VCR works at all after a fire, failure to remove the layer of soot in its interior might cause considerable damage later.

If your VCR survives a fire, clean it (and any other electronic gadgets for that matter) as quickly as possible. This minimizes smoke damage to the exterior cabinet. Thoroughly wipe the outside of the VCR using a damp sponge. This will pick up the bulk of the soot.

Matters are a little different on the inside of the deck. Using a household cleaning spray on the mechanical and electronic parts of a VCR is not recommended, partly because the cleaner is water-based and poses a potential short circuit hazard, and partly because it is not really effective in cleaning hard-to-reach places.

A cleaner/degreaser spray can be used to thoroughly clean the inside of the deck. Spray the cleaner heavily to remove the soot. Use the extension spray tube (usually included with the can) to get at hard-to-reach places. The cleaner evaporates after 5 or 10 seconds. If soot remains, spray again. If the smoke build up is heavy, spray the cleaner on and brush it off with a small painter's brush.

The cleaner/degreaser leaves no residue, but it can remove the lubrication on some of the mechanical parts in the deck (including the threading posts and rotating parts). After using the cleaner/degreaser, lubricate the deck, if necessary, following the instructions provided in Chapter 5. But remember that VCRs, on the whole, require little lubrication.

If you're not sure if a part needs oiling or greasing, try using the deck for a while. Watch its operation carefully to see if lubrication is called for. You'll have to leave the top cover of the VCR off to do this.

Finally the heads, including the rotating video heads, will need a complete cleansing as well. Follow the cleaning procedures discussed in Chapter 5.

WATER DAMAGE

Each year, water does more damage to personal property than fire. Even if you don't live in a flood basin, there is always a chance that your VCR can be subjected to a liquid drink. If you have a camcorder, you could drop it into the water while lounging near the pool. It's happened before, and it can happen again.

If your VCR is ac operated and it becomes wet, *immediately* unplug it. If it is unsafe to reach the plug, turn the power off at the circuit breaker or fuse box. Do not touch a wet VCR; you might receive a bad shock. If the deck shorts out when water spreads inside its cabinet, there is a very good chance that it is seriously damaged and should be taken to a repair shop for proper service. If there is no sign of immediate short circuit damage, you can minimize any further problems by following the steps below.

Removal of Excess Liquid

Of first importance is to soak up the excess liquid. Use paper towels to blot up the extra. If you feel any liquid has seeped into the machine, you must disassemble it and remove the standing water from the inside as well.

If the deck was dunked into fresh water, you need only to wait until the remaining moisture evaporates. Some water could be trapped under components, so even if the surface of the circuit board and internal parts are dry, there still might be water lurking underneath. You may use a hair dryer—on low or no heat only—to help speed up the drying process. Do not dry with high heat, as you might warp some parts. Even after the water is gone, some moisture might still be present, particularly on smooth metal parts and around the head drum area. Allow another two to three hours for condensation to evaporate.

You can test for remaining moisture by placing the VCR in a plastic garbage bag. Seal the end and place the bag overnight in a warm but dry place. If there is moisture remaining in the deck, condensation will form inside the bag. It's a clue to let the machine dry out some more.

Some decks have a built-in dew circuit that prevents playback if there is condensation in the unit (older decks had a dew light that turned on to indicate excessive moisture). If your VCR still doesn't work after drying, this might be the problem. Pay particular attention to the head drum. Condensation on the polished surfaces can cause the tape to stick and jam or stop, and may cause damage. If the deck turns on but won't play a tape, keep it on for a while to help burn away the moisture.

Removing Sticky or Staining Liquids

If the VCR was subjected to sticky or staining liquids—such as salt water, coffee, sugary soda pop, and milk—you need to thoroughly clean the deck inside and out. This prevents residue from the liquid from interfering with the operation of the unit (salt water will corrode the metal parts, for example). If the liquid is thick, use a damp sponge to wipe up as much as possible.

Next, thoroughly spray the VCR with a cleaner/degreaser, as mentioned in the section above on fire damage. Apply the cleaner/degreaser heavily until all signs of the residue are gone.

Because the cleaner/degreaser is non-water-based, you can also use it to remove any water that may remain in the deck. The cleaner/degreaser will act to displace the water, removing it from even hard-to-reach places.

The remote control transmitter is often subjected to heavy abuse, because it can be placed on tables along with food and drink. Crumbs, liquids, and other matter can easily fall in the cracks and muck up the switch contacts inside or short out wire terminals and connections. Disassemble the remote, as explained in Chapter 5, and thoroughly clean the switches. You can spray the cleaner on or use a damp cotton swab, as shown in Fig. 6-4.

After cleaning, moving parts might require lubrication. See Chapter 5 for more details on lubricating VCRs. Be sure to clean all the magnetic heads prior to use.

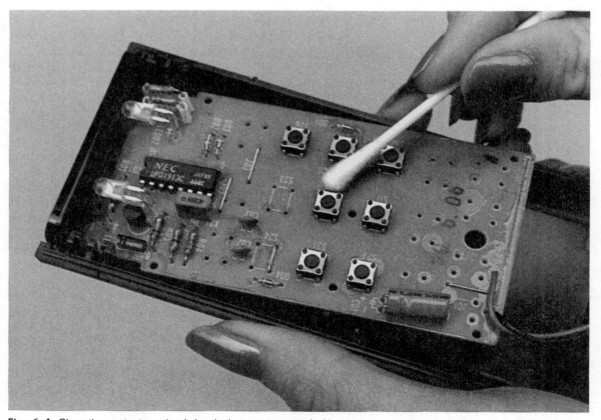

Fig. 6-4. Clean the contacts and switches in the remote control with a cotton or sponge swab dipped in suitable cleaner.

After Cleaning Test

When you are satisfied that all the moisture has been removed, and that the deck has been properly lubricated and cleaned, perform a leakage current test as explained earlier in this chapter. If the leakage test proves negative, plug in the unit and turn it on. Test the VCR completely for proper operation. Should you find that some moisture is still present inside the VCR, stop testing immediately and allow the deck to dry out more.

SAND, DIRT, AND DUST

The video heads in a VCR are extremely sensitive to even a small amount of dirt. Decks used outside in dusty or sandy environments are susceptible to premature failure, especially if they are accidentally dropped in the sand at the beach or carried on a long-distance desert hike, where the dust from the great outdoors permeates everywhere.

Removal of every last speck of sand, dirt, and dust is crucial to ensure proper operation of the VCR and to prevent possible damage to the video heads, video head drum, and tapes. This is particularly true if the deck is filled with gritty sand. A soft brush can be used to wipe away the bulk of the sand, or if there is lots of it, you can use a vacuum cleaner with a brush attachment. Remaining sand, dirt, and dust can be removed from the cabinet, chassis, and interior with a tac cloth designed for wiping away sawdust from woodworking projects. Tac cloths are available at any hardware store.

Even though the visible dirt and sand might be gone, there still could be an accumulation of micro-sized particles of dust. You can clean the entire inside of the VCR by liberally squirting it with your trusty can of cleaner/degreaser. The fluid in this cleaner evaporates quickly, leaving no residue. The cleaner/degreaser is a solvent, but it is not petroleum based so it will not harm metal or rubber parts.

Sand and dirt have a way of getting into the most unusual places, and it might take several hours to displace it all. Be sure to inspect the input and output jacks on the deck. Grit inside will prevent proper electrical contact, so you might not get any sound or picture. You can clean the jacks with a swab soaked in alcohol (see Fig. 6-5). Clean both the inside and outside.

As mentioned before, you might need to oil and lubricate the spinning and sliding parts in the deck after cleaning. Follow the directions outlined in the Chapters 4 and 5, and be sure to go lightly on the oil and grease.

FOREIGN OBJECTS

Just about any service technician of audio and video equipment will tell you that a good portion of machine repairs are caused by people inserting foreign objects into the mechanism. Horror stories of this sort abound, and most are innocent mistakes caused by children. How about the one where a five-year-old boy stuck a peanut butter sandwich into the tape loading drawer because he wanted to know what kind of picture it would make? Or the time when a young girl used her father's VCR as

Fig. 6-5. The input and output terminals can be cleaned using a cotton swab dipped in alcohol or non-petroleum-based solvent cleaner.

a piggy bank? She put all her extra pennies, nickels, dimes, and quarters into the ventilation slots of the deck. She had amassed quite a savings until the VCR suddenly went on the fritz.

This kind of accident is best avoided by warning young children that the VCR is not a toy. You can make your own rules in your home, but if you allow your children to use or even touch the deck, spend a few moments instructing them on the right and wrong way to play tapes. Children are naturally curious about things, especially something new like a VCR, and you'll greatly reduce the chance of a serious accident if you provide adequate do's and don'ts ahead of time.

Despite the best efforts, warnings, and rules on your part, foreign objects might still get lodged inside the machine. Some might even be your fault. When this happens, immediately turn off the deck and unplug it. If the object cannot be completely retrieved, disassemble the VCR and remove it.

Things like sandwiches, cookies, candy, and other objects can often leave their mark even after you have removed them! You can use a brush or vacuum to remove the crumbs. Heavier or stickier sediments will require removal with a squirt from the cleaner/degreaser.

LEAKED BATTERIES

Portable VCRs and camcorders, as well as remote controls for home decks, run off battery power. Given the right set of circumstances, even the best-made battery can leak. The acid from the battery oozes everywhere and not only blocks electric current, causing failure, but corrodes the inside of the battery compartment and the battery terminals.

Your best defense against leaking batteries is to remove them if your deck or remote will not be used for a long time. Batteries tend to leak the most when they run out of electrical juice, so a battery that sits unused in a portable or remote has a very good chance of leaking. If the batteries are still good when you take them out (you can test them with a battery tester or volt-ohmmeter), use them in something else.

When storing batteries—new or used—keep them in a cool, dry place. You can greatly prolong the shelf life of batteries by storing them in the refrigerator (not freezer). Again, batteries can leak just about everywhere for any reason, so to prevent contaminating your food with battery acid, wrap them up in a sealable food storage bag.

Removing Battery Acid

Should the batteries leak in your camcorder, portable, or remote, remove them immediately and throw them away. Avoid excessive contact with the battery acid, as it can burn skin. Use a lightly dampened cloth to remove the excess battery acid deposits from the battery compartment. If the batteries leaked only a little bit, there might not be any deposits in the compartment or on the terminals. If the terminals look clean and bright, install a new set of batteries and test the unit.

If there are excessive battery acid deposits, clean the entire compartment with isopropyl alcohol or with the can of cleaner/degreaser. Use a cotton swab to scrub the cleaner onto the terminals. Also, you can remove any remaining battery acid deposits by rubbing the surface of the contacts with a pencil eraser.

7

Troubleshooting and Repairing Non-VCR Problems

If you suspect that any part of your VCR isn't working properly, take a few moments and follow the simple steps outlined in this chapter before tearing the deck apart. In most cases, the problem might not be a malfunction at all. Quite often, the nasty culprit will be a maladjusted TV set, a loose or broken wire, or a bad videotape.

START AT THE TV

Start at the TV and work your way back to the VCR. Consider the tuner, settings, picture controls, and reception problems.

Proper Channel and Fine Tuning

If you are not getting a picture from your VCR, check the controls on the TV first. One of the most common problems is that the TV is not tuned to the proper output channel. Most all VCRs output to either channel 3 or 4.

If your TV has a fine-tuning control, adjust the tuning knob on the set as you play back a tape on the VCR. The channel frequency generated by your video equipment could be somewhat different than what is normally broadcast by a TV or cable station, so you might have to make fine-tuning adjustments to get a crystal clear picture.

Proper Switch Selector

If your TV is connected to a video selector switch, be sure that it is dialed to VCR and not to the antenna, cable, or cable decoder box.

Vertical Hold Control

If the picture on the TV rolls when a pre-recorded tape is being played, check the vertical hold control on your set. If your TV lacks a vertical hold control or it isn't accessible, you might need to purchase a video stabilizer to combat the effect of the copy protection recorded on the tape (see Chapter 3, "The VCR Environment," for more details on video stabilizers).

Brightness, Contrast, and Other Controls

Should the picture be dark or washed out, try adjusting the brightness and contrast controls. Colors that seem "off" could be the result of maladjusted color and tint controls on the TV. If your set has auto-color circuitry, be sure it is switched on.

Isolating Reception Problems

To isolate a reception problem on your TV, hook up your cable or antenna directly to the set, bypassing everything else in your system. Take off the connection from an rf modulator, VCR, or computer, and even remove the direct audio and video connections if these have been made. If reception improves, the problem lies in an accessory coupled to the TV or the cabling that connects everything together. Reconnect each piece of equipment and its cabling, in turn, to see which one is causing the problem.

Flickering Picture

If you see a flickering picture when viewing a pre-recorded tape, you might be witnessing Macrovision anti-copying signals interfering with your TV. If you have an automatic gain control (AGC) switch on the TV, switch it off. If that doesn't help, try the fine-tuning knob.

The flickering picture might also occur if you have two VCRs connected to one another, even if you aren't actually recording. Depending on how you have your system connected, the Macrovision signals could be interfering with the AGC circuits in the second VCR. Try disconnecting the VCRs and viewing the tape directly from one VCR to the TV, as shown in Fig. 7-1.

If all else fails, a video stabilizer built to counter the Macrovision signal encoding might be required. Consult your local video dealer. Sellers of video stabilizers often advertise in the video magazines. Flip through a current issue and note the ads.

PROGRAM RECEPTION

If the picture you record on your VCR isn't what you think it ought to be, it might not be dirty heads or a bad tape. The program reception, either through the antenna or cable system, could be at fault.

General

To effectively track down the cause of persistent ghosts or interference, hook up your TV directly to the incoming signal source—either the cable or antenna—

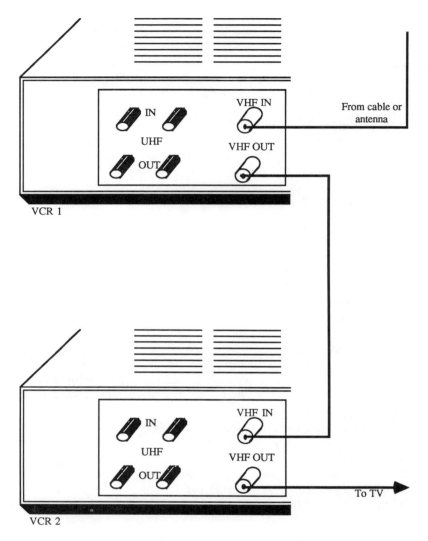

Fig. 7-1. A) Improper connection of VCRs when watching Macrovision-encoded tapes.

thus bypassing your VCR and other components in your system. If it's truly a reception problem, the interference will remain.

Check the cable(s) that connect to your TV. If you are viewing tapes from your antenna, be sure the antenna is still in good shape and that it is pointing in the proper direction. Double-check the condition of the cables leading to the television. Inspect

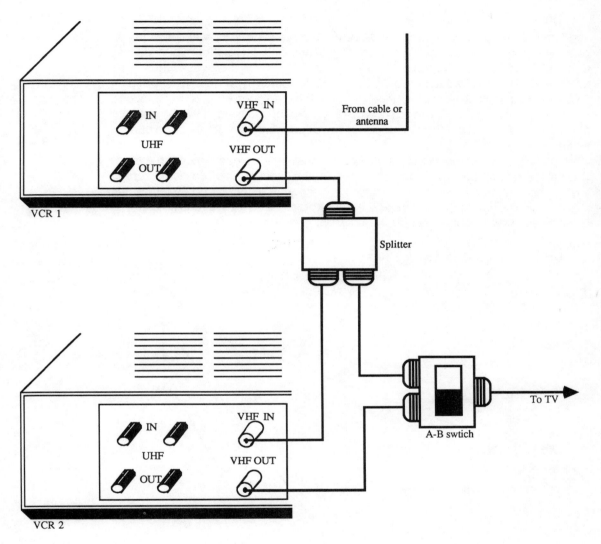

Fig. 7-1. B) Proper connection using an A/B switch.

the cabling and all in-line components (matching transformers, splitters, etc.) from the antenna to the TV set.

If you're hooked up to cable, call your local cable company and have them troubleshoot their system.

TROUBLESHOOTING AND REPAIRING NON-VCR PROBLEMS

Ghosts

Ghosts are created by the TV signal reflecting off water, water towers, tall buildings, or other structures. Your antenna receives both the original signal directly from the TV station and the reflected signal(s) that produces the ghost. Ghosts can come and go in bad weather, so wait out a storm before you do anything harsh. If you find you can't eliminate the ghost, try aiming the antenna in different directions, even away from the TV station, as shown in Fig. 7-2. The main signal will be reduced and so will the ghost, but the ghost should be reduced much more than the original signal.

Of course, if the signals from the other channels are not coming from the same source, then the reception on other channels may be affected. You can also purchase a ghost eliminator, a small device that fits between your antenna and TV. However, a ghost eliminator might not work well if the ghost signal is exceptionally strong. A more directional antenna may be the solution.

Dots, Dashes, and Wavy Lines

Sputtering interference—dashes and dots across the screen and a buzzing in the sound—usually points to an electrical disorder. Common causes are vacuum

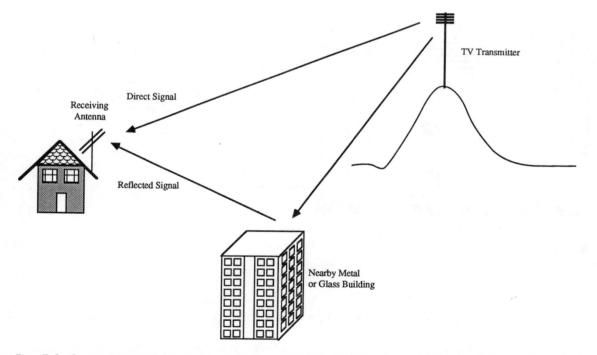

Fig. 7-2. Ghosts are usually caused by receiving most of the direct and some of the reflected signal from the TV transmitter. The ghosts can be reduced or eliminated by carefully positioning the antenna or using an antenna that is highly directional.

cleaners, car engines, hair dryers, and neon lights. If the problem is caused by an electrical appliance and it's bothersome enough, try installing an in-line power filter. These attach to your set's ac cord. If the problem is caused by radio interference like a car engine, get an antenna filter. These attach between your TV and antenna.

Interference in the form of wavy lines can be caused by such things as cordless phones, CB and ham radio transmitters, FM stations, automatic garage door openers, and other devices that emit radio frequency signals. Again, a filter (or trap) attached to the vhf terminals on your set the TV and antenna or cable outlet usually helps. Traps are often available to combat a particular kind of interference, so you might have to experiment before you lick the problem. Such traps or filters are available from Radio Shack and other electronics outlets.

VCR CABLES AND CONTROLS

Like the TV, problems with a VCR are many times caused by improper connections and maladjusted controls. Make sure all of the cables are attached properly and that each one is connected to the right input or output. Double-check the cable connections by referring to your operator's manual, especially after disconnecting your VCR to move or clean it.

Go through the same procedure outlined above for making sure you are on the proper channel. If you are connected to cable, the decoder or converter box might have a channel 3 or 4 output, so you must dial your VCR to that channel to receive a picture. Also check the TV/VCR switch. This switch allows selection between VCR playback and over-the-air program viewing through the VCRs tuner.

TRACKING CONTROL

The tracking control on the VCR helps assure synchronization with tapes made on other machines. If the picture is snowy, particularly on the top or bottom of the screen, adjust the tracking control. In some cases, the control will have little or no effect. But just because the control doesn't work doesn't mean that the VCR is at fault. In nine times out of ten, the tape will be faulty or recorded on a deck that is poorly aligned. You can isolate the problem to the VCR by trying other tapes in your collection.

TV/VCR SWITCH

VCRs have a TV/VCR switch that lets you manually switch between over-the-air viewing through the VCR and tape viewing. When in TV position, you see the channels on your TV as they are received by the tuner in the VCR. Your set is tuned to the output channel of the VCR, usually channel 3 or 4. In VCR position, you see the playback of the tape.

Depending on how you have your VCR connected to your TV and video system, pressing the TV/VCR switch when watching a tape may switch you to over-the-air viewing. In older models, the position of this switch is not reset each time you press the PLAY button of your VCR. So even though your VCR is working properly, your

TROUBLESHOOTING AND REPAIRING NON-VCR PROBLEMS

TV will continue to receive the broadcast channel through the VCR's tuner, not the program on the tape. The moral: Always check the position of the TV/VCR switch. When you are not getting a picture, push it to see if the taped program comes on.

REMOTE INTERFERENCE

Let's say you are trying to play a tape, but pushing the buttons on the VCR has no effect. Bad tape? Maybe. Bad VCR? Probably not. What else may cause the problem? The remote control. If your VCR has a remote control, either wired or wireless, its controls usually override the main controls on the front panel of the deck. Locate the control and make sure that no books, plates, magazines, or other debris is pushing on the buttons, and improperly commanding the VCR.

In some cases, a button on the remote is stuck, so even though the controller seems healthy on the outside, it is still interfering with the operation of the VCR. If the remote is the wired type, disconnect it from the deck. If the remote is the wireless type, take it into another room, remove its batteries, or cover up the front to block the passage of infrared light.

USING A KNOWN, GOOD TAPE

If your VCR is suffering from one or more picture and sound problems when playing tapes, test the deck with a known, good tape. Always do this before passing judgment on your VCR. A known, good tape is one you've used recently that seemed okay, or a relatively new store-bought pre-recorded tape.

There are two general forms of out-of-order cassettes: broken cassette shells and damaged tape.

Broken Cassette Shells

A broken cassette shell is one that is cracked or internally damaged so that the tape no longer moves through it smoothly. A broken roller might prevent the tape from coming out, and this will cause the tape to jam inside the cassette. Similar problems occur when the cassette hatch (the door on the front of the shell) snags with the tape, or when the internal reel locks inside the shell fail to disengage (for a complete description of the inside of VHS, VHS-C, Beta, and 8mm cassettes, see Chapter 2). You might be able to repair the cassette by moving the tape to another shell, as discussed in Chapter 5.

Damaged Tapes

A damaged tape is rarely repairable. By damaged, we mean a tape that has been folded, spindled, or mutilated in such a way that the VCR can no longer read the signals impressed on it. A stretched tape, as shown in Fig. 7-3, is not only dimensionally thinner than a good tape, but it buckles and prevents the video, audio, and control track heads from reading the information.

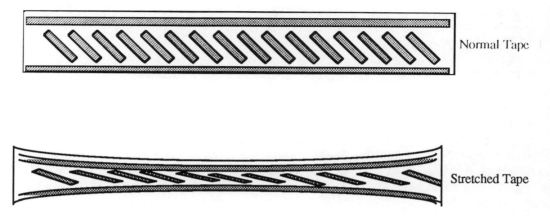

Normal Tape

Stretched Tape

Fig. 7-3. Stretched tape deforms the signal area and impairs playback.

Tape edges that have been cut or scored can cause audio or tracking problems. In all formats, the linear audio tracks are along the top of the tape and the control track is along the bottom.

A problem with the tape shell or VCR can scratch the tape. If the problem is serious enough, the scratch can remain for the entire length of the tape (much like a scratch in film caused by a dirty projector). The scratch shows up as a streak of white or black on the screen, and can cause the video heads to become prematurely clogged.

A tape that has spilled out of the cassette, either inside or outside of the VCR, is subject to mutilation. At the very least, the tape will be folded and creased along its length. You should not play a tape damaged in this manner because it can actually snap off a video head. If the creasing is light, you can use the tape, but it's better to make a copy and throw the original away.

Another alternative is splicing out the bad portion. But unless the splice is very well made (as discussed in Chapter 5), you run the risk of damage to the video heads. The reason: the video heads don't just skim across the tape, they actually protrude and penetrate into the tape, causing it to bulge out. A gap in the splice is like a pothole for a car—the head can dig into the gap and break off. Tape jamming and spillage can occur if the tape ends are incorrectly aligned, as shown in Fig. 7-4.

TROUBLESHOOTING CHECKLIST

To recap, before you tear open your VCR, check these components of your video system first:

- ✦ TV operation (including all controls).
- ✦ Anti-copying signal on pre-recorded tape.
- ✦ VCR-to-TV cables.
- ✦ Antenna or cable system hookup.

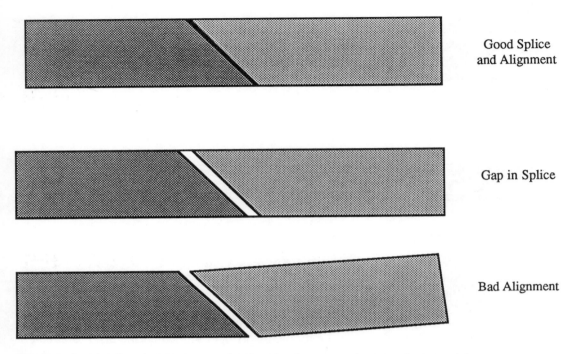

Good Splice
and Alignment

Gap in Splice

Bad Alignment

Fig. 7-4. Avoid splices that are improperly aligned or leave gaps between the tape ends.

- ✦ Reception.
- ✦ TV/VCR switch.
- ✦ Tracking control.
- ✦ Remote interference.
- ✦ Tape quality.

8

Troubleshooting Techniques and Procedures

The word "troubleshooting" literally means aiming at trouble and firing away until you hit the bulls-eye. In a more practical sense, troubleshooting means locating and eliminating sources of problems but doing so in a logical and predescribed manner.

Troubleshooting is the basis of electronic and mechanical repair, and a thorough grasp of its techniques and procedures is important. Just as you can't hope to shoot at a target with a blindfold over your eyes, you cannot wildly attack a problem in your video cassette recorder and pray that you luck onto the solution. Troubleshooting lets you approach the problem from all angles and zero in on the cause with the least amount of wasted time, energy—and most of all—money.

This chapter details the concepts behind VCR troubleshooting techniques and how to apply them to actual hands-on procedures. If you are already familiar with electronic and electromechanical troubleshooting, the information in this chapter may seem old hat to you. If so, skip to the next chapter. It contains troubleshooting flowcharts that you can use to pinpoint common problems with your VCR. If you are not already familiar with basic troubleshooting techniques and procedures, be sure to read this chapter, as it contains useful information you won't want to miss.

THE ESSENCE OF TROUBLESHOOTING

You can better understand the role that troubleshooting plays in the repair of VCR ailments by using a more familiar concept: figuring out what's wrong with the family car. Let's say you go out to your car one morning, turn the key, and the car won't start. The battery turns the engine, but even after 10 or 15 tries, the car simply

TROUBLESHOOTING TECHNIQUES AND PROCEDURES

won't start, and every minute you spend cranking the engine is every minute you are late for work.

You could open the hood, tear out the engine and rebuild it, or you could replace random parts, thinking that it's got to be one of them that's causing the problem. You know better of course, and you stop and think for a minute: why won't the car start? The engine turns over, so you can rule out a bad battery. But the problem could be in the ignition system, the fuel lines, or a number of other sub-systems. In fact, there are several possible causes:

- ◆ The engine is flooded.
- ◆ The spark plugs are fouled.
- ◆ The engine isn't getting gas.
- ◆ The engine timing is off.
- ◆ The spark isn't strong enough.
- ◆ There is no spark.
- ◆ The engine isn't getting enough air because the choke is closed.

. . . And so forth. Once you have identified the possible causes, you can take corrective steps on a one-by-one basis. You start with the most common or probable cause and work your way down. In most cases, failure to start the engine is caused by flooding. To remedy this, you open the hood and remove the air filter and perhaps a couple of spark plugs. The gas evaporates and you start your car. If the problem still persists, you go to step two, and so forth.

We've just outlined the basic three-step process to troubleshooting, and it applies to VCRs just as it does to hard-starting cars.

Step 1: Analyze the symptom ("car won't start but engine turns over") and develop a list of possible causes.
Step 2: Arrange the causes in order, from the most likely to the least likely.
Step 3: Start at the top of the list (most likely cause), and by a process of elimination, inspect, test, or otherwise rule out each possibility until the problem is located.

Once the problem has been found, you can clean or repair the faulty component.

TROUBLESHOOTING FLOWCHARTS

It's sometimes easier to visualize the troubleshooting process by using flowcharts. These are graphic representations of the possible causes and their solutions. The exact form of the flowchart can vary, but the basic information they contain is the same.

One example of a troubleshooting flowchart is shown in Chart 8-1. The chart is labeled with the problem (or symptom), which in this case is "Engine Turns Over But Won't Start." Below the title is a set of boxes. The boxes are stacked vertically, and each box on the left hand side contains a possible cause for the malfunction, usually a bad or dirty part or sub-system. The boxes are organized from the most

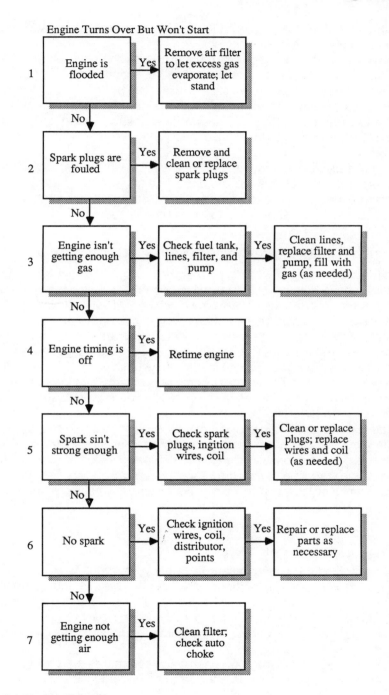

Engine Turns Over But Won't Start

1. **Engine is flooded** — Yes → Remove air filter to let excess gas evaporate; let stand

2. **Spark plugs are fouled** — Yes → Remove and clean or replace spark plugs

3. **Engine isn't getting enough gas** — Yes → Check fuel tank, lines, filter, and pump — Yes → Clean lines, replace filter and pump, fill with gas (as needed)

4. **Engine timing is off** — Yes → Retime engine

5. **Spark sin't strong enough** — Yes → Check spark plugs, ingition wires, coil — Yes → Clean or replace plugs; replace wires and coil (as needed)

6. **No spark** — Yes → Check ignition wires, coil, distributor, points — Yes → Repair or replace parts as necessary

7. **Engine not getting enough air** — Yes → Clean filter; check auto choke

Chart 8-1 Flowchart for engine trouble diagnosis.

TROUBLESHOOTING TECHNIQUES AND PROCEDURES

likely to the least likely, so you won't waste your time with an unusual fault when the actual cause is really quite common.

In each level are two or three boxes. These boxes contain the suggested remedy, and in some cases, the procedure for testing the suspected part or sub-system.

Arrows connect the boxes in such a way that by thinking of the boxes as questions, you can answer them simply with a Yes or No response, and navigate yourself around the chart. Each No answer to a possible cause brings you down to the next level. Answer Yes to a cause, and you move laterally as you make tests and repairs.

In actual practice, however, you don't really know where the problem originates until you test each possible cause. Therefore, at each level you initially answer with a Yes response, until you have the opportunity to test it out. If the test proves negative (not the cause), you move down to the next level.

USING THE TROUBLESHOOTING CHARTS IN CHAPTER 9

The troubleshooting flowcharts in Chapter 9 follow the same form and logic as the example above. If your VCR isn't working properly, find the chart that most closely matches the symptoms of the deck. Some machines exhibit multiple symptoms, and you might need to consult several charts as you attempt to pinpoint the cause.

Once you've identified the proper chart to use, start at the top level and work your way down, eliminating those causes that you're sure are not the source of the problem. Double-check your work to make sure you haven't missed something, and don't forget the obvious. By far, the majority of problems with VCRs are caused by seemingly innocent things, like an old tape, a dirty head, a dirty switch, or a selector on the deck or amplifier that's not set correctly.

Whatever you do, avoid the temptation to tear your VCR apart before you adequately identify the cause. Disassembly and removal of any part, even the top cover, should only be done until after you've eliminated the other causes. Remember that some of the components in VCRs cannot be repaired or replaced without special tools and techniques, so make it a point to not remove any parts unless you absolutely have to.

WHAT YOU CAN AND CAN'T FIX

You can try to fix anything in your VCR. It's yours, you own it, and you're free to do anything you like with it. Many problems are easily handled by the home technician, and there is no reason why you should not diagnose and repair them yourself. However, there are a number of malfunctions that are best left to a repair technician who has access to the service literature and special alignment and test jigs for your specific deck.

Here is a general list of the components and assemblies you can fix in your VCR yourself, and those that should be referred to a qualified technician. This list is not absolute by any means. Your level of expertise and the amount of service material and tools you have or can obtain greatly influence the types of repairs you can perform.

What You Can Do:
- ✦ Clean and inspect the video heads.
- ✦ Clean and inspect the stationary erase, audio, and control track heads.
- ✦ Clean the exterior and interior to remove dust, nicotine, and other contaminants.
- ✦ Clean and replace belts and rubber rollers.
- ✦ Inspect and replace springs.
- ✦ Lubricate gears, shafts, and other moving parts to prevent them from grinding.
- ✦ Replace or resolder broken wires.
- ✦ Clean or replace dirty or broken switches, including front panel switches.
- ✦ Test for proper operation of the capstan, take-up reel, and loading mechanism motors, and replace them if necessary (these usually do not require critical alignment).
- ✦ Replace the main printed circuit board (PCB) and ancillary boards (special handling of the boards required).
- ✦ Replace common electronic components (resistors, capacitors, diodes) on the PCBs, but only if the exact original value is known.

What You Should Not Do:
- ✦ Disassemble the video head drum.
- ✦ Replace video heads.
- ✦ Adjust the heads and guide rollers inside the deck.
- ✦ Adjust trimmer potentiometers on the main PCB, without the use of a schematic and an oscilloscope.
- ✦ Replace integrated circuits, transistors, or any other component where an exact replacement or substitute cannot be guaranteed.

WHAT TO DO IF YOU CAN'T FIX IT

Should you find that repairing your VCR is beyond your technical expertise and resources, don't fret. Make a note of the problem and the suspected cause, reassemble the deck, and take it to be serviced. By performing the basic troubleshooting procedures yourself ahead of time, you make the repair tech's job that much easier. You might not get a reduction in the labor charge, but if the problem is fixed properly the first time around, you won't have added any costs due to "additional work" the technician had to do because the machine was opened. A VCR that must be returned to the service shop time and again is often the result of a communications problem between you and the repair technician.

Some repair techs and service centers frown on consumers repairing their own VCRs. But if you've followed the directions in this book and in the manufacturer's literature and used the right tools and cleaning supplies, you've done nothing that the service shop would not have done—and charged you handsomely in the process. When performed properly, routine maintenance and troubleshooting procedures do not harm the VCR.

Avoid a repair shop that adds an additional service fee simply because you've opened the top cover and made some preliminary tests. Unless you have broken the deck even more, the extra service fee is completely unnecessary and very unethical. After all, an automotive garage would not add $50 to their price of a tune-up because you changed the oil.

If you are sure that the fault lies in a component that cannot be repaired and replacement does not require critical adjustment, contact the manufacturer and ask for a parts list (if you don't already have it). Identify the faulty component and order a replacement. Most manufacturers that sell replacement parts to the public ask for payment in advance, and the easiest way to do this is by credit card. Few will ship COD.

Let's say the manufacturer proves uncooperative—what then? Try to obtain the part from a repair center authorized to work on your brand of player (in this case, "authorized" simply means that the repair center has an open channel for replacement parts). The shop may tack on a small service fee, which is understandable.

Receiving the replacement parts from the manufacturer or a repair center might take as little as a week or as long as several months. At worse, the broken part must be ordered from some warehouse in Japan, and it seems as if they ship it to the U.S. by sailboat. You've heard of the slow boat to China; in the case of replacement parts for VCRs, it's the very slow sailboat from Japan. Nevertheless, owning an American- or European-made player is no guarantee that parts will arrive any sooner.

Keep records of when you ordered the part and whether you've written any follow-up letters or phoned the manufacturer directly. Ask for the names of everyone you speak with and write them down. You never know when this additional information will come in handy.

TROUBLESHOOTING TECHNIQUES

Effective troubleshooting depends mostly on common sense, but here are some tips you'll want to remember.

Write Notes

Write copious notes. Write everything down, including how you removed the top cover, volt-ohmmeter readings you made, visual observations, parts you replaced—in short, anything and everything. By keeping notes, you will not only be able to retrace your steps should you get hung up on a particular fault, but you'll be able to better deal with recurring problems. The maintenance log in Appendix C provides blanks for notes; use additional sheets of blank paper if necessary.

Use the Proper Tools

Refer to Chapter 4 for more information on the proper tools to maintain and service your VCR. Don't make do with a tool that was not designed for the job. If you don't have the required tools and supplies already, spend a little extra on them. The maintenance and troubleshooting procedures outlined in this book require only the most basic hand tools and test equipment. Expensive items like oscilloscopes, function generators, and frequency counters are not absolutely required unless you opt for more detailed troubleshooting of your own.

Use Your Volt-Ohmmeter Correctly

A number of troubleshooting procedures require you to test the suspected component or assembly with a volt-ohmmeter. Be sure to use this piece of equipment correctly, or you could wind up missing a potential fault, or replacing components that are perfectly good.

To test continuity, select the resistance function on the meter. If the meter is not auto-ranging, choose an initially high resistance range, on the order of 10 KΩ or more. Attach the two meter leads (or probes) to either side of the switch, wire circuit, or connector you are testing, such as that shown in Figs. 8-1 and 8-2. Double-pole, double-throw switches, sometimes used in power switches, have extra terminals to check. Figure 8-3 shows how to use your meter to test one of these switches.

For the most part, continuity is a "go/no-go" test. You will either get a reading of 0 ohms, which means that the circuit or connection between the two test leads is complete; or infinite ohms, which means that the circuit is broken or that the connection between the two test leads is open.

Sometimes, a reading of 0 ohms is exactly what you want, like when you are testing the closure of a switch or a length of wire to make sure it is not broken inside. Other times, 0 resistance means that something is shorted out, which is an unhappy

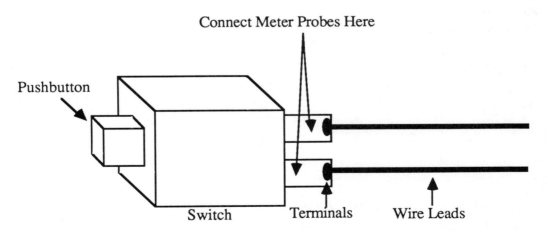

Fig. 8-1. Connect the meter probes to the switch terminals to test for proper operation of the switch.

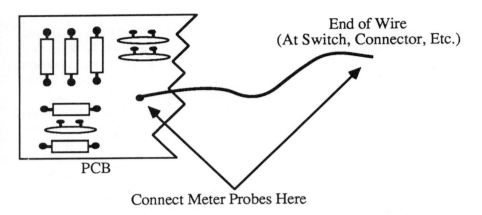

Fig. 8-2. Check the continuity of a wire by attaching the meter probes to both ends of the wire.

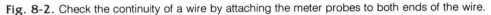

situation. Likewise, a reading of infinite ohms (which meters display in a variety of ways), may be correct for a given test; 0 ohms for another test. The troubleshooting charts in Chapter 9 provide more details on the typical readings you should get for any given instance.

When using the meter, make sure that you do not touch the metal part of the probes. Your body has a natural resistance, however high, and you will add or subtract it to the measurement you're trying to take. Always take your readings by grasping the test leads by the plastic insulator.

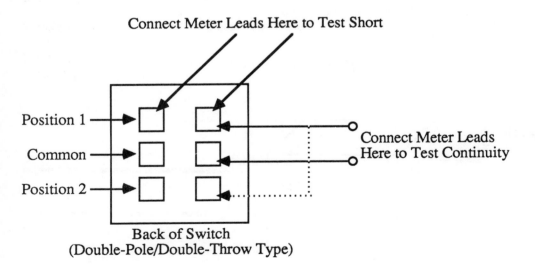

Fig. 8-3. The correct test points when checking a DPDT switch.

Volt-ohm meters test more than continuity, of course. For example, you will want to use the ac and dc voltage functions to test for proper voltage levels going into and out of the deck's power supply. Be sure to select the correct function *before* you connect the test leads to the live circuit. With many meters, connecting the leads to a dc voltage when the unit is dialed to ac can burn out a fuse or cause internal damage.

If the meter is not the auto-ranging type, always choose a range higher than the input voltage. If you are not sure of the voltage level, choose the highest one first, then work down. Most all digital meters have built-in overload protection to prevent damage by choosing a two-low range. However, the needle in an analog meter can be permanently bent or broken if you accidentally choose a range that's too low, and the needle violently swings to the end of the scale (''pegs'').

GOING BEYOND THE FLOWCHARTS

Few video cassette recorders are designed exactly alike. Differences between brands, even between models of the same brand, require slightly different troubleshooting procedures. Use the flowcharts in Chapter 9 as a starting point only. The design and construction of your VCR might require additional troubleshooting steps. If possible, refer to the schematic or service manual of your machine. It will often detail testing points or procedures that are unique to the particular model.

In other cases, the design of your VCR might not require that you perform some troubleshooting procedure. For instance, if your deck is a top-load type, it lacks a motorized tape loading mechanism, so problems with this assembly are of no concern to you. Likewise, camcorders will lack any type of internal power transformer, so it is not necessary to test it if you are having trouble with powering your unit (except for the battery or ac adapter).

To keep the charts as simple as possible, we've limited them to reflect problems inherent in front- and top-loading ac-operated home models. If you own a portable VCR or camcorder, allow for the design difference in your troubleshooting procedures. Keep in mind that portable decks and camcorders are considerably more difficult to work with because of the miniaturization of scale.

9

Troubleshooting
VCR Malfunctions

VCRs are complex electronic and mechanical devices. Their reliance on special proprietary integrated circuits and zero-tolerance mechanical alignment leave the average consumer or electronics hobbyist limited opportunity for complete home repair.

This does not mean, however, that doctoring the ills that beset the average VCR is entirely out of your hands. On the contrary, the new modularity of VCR design makes mechanical and electronic faults very rare—assuming that the machine was well manufactured in the first place.

In this chapter, you'll learn what you can do when your VCR goes on the fritz and how to go about fixing it. The emphasis is on troubleshooting the mechanical sections of the deck, and plenty of flowcharts are provided to graphically show the step-by-step process of the cure for a given ailment.

FIX IT YOURSELF

By far, the most cited problems with video cassette recorders are not internal problems with a transistor or IC chip or even a broken video head. Most of the problems are relatively mundane mechanical faults caused by failure to follow the operating instructions, careless use or abuse, tape defects or damage, clogged (or dirty) video heads, and just normal wear and tear.

More specifically, service shops for VCRs report that the following causes represent by far the greatest majority—over 85 percent—of warranty and non-warranty repair. The causes are listed from most common to least common.

- ✦ Poor system connection (cable from VCR to TV, etc.).
- ✦ Improperly set controls (especially control knob).
- ✦ Dirty heads (video, audio, and control track).
- ✦ Tape problems (includes bad or old tapes).
- ✦ Physical abuse or accidental damage (caused by neglect or improper cleaning).
- ✦ Dirty, glazed, or cracked reel spindle idler tire.
- ✦ Dirty, glazed or cracked drive belts.
- ✦ Jammed mechanical components due to oxidation, dirt, or loss of lubrication.

As you can see, you can save yourself the cost of a service call, not to mention headaches and frustration, by not only following manufacturers' instructions for operating your video gear but by caring for your tapes and your deck with routine maintenance, as discussed in Chapter 4.

SERVICE POLITICS

Inevitably, ordinary mechanical and electronic breakdown does occur. You can repair a number of these breakdowns yourself. Most dysfunctions can be corrected by cleaning dirty electronic contacts and switches, oiling or greasing the moving parts, repairing the occasional broken or frayed wire, and replacing worn parts like rollers, belts, and plastic gears.

Unless you have specific training and experience in working with both high-speed digital and analog circuits and have the proper servicing tools, problems with the video heads (other than dirt), as well as failed integrated circuits and other components on the circuit boards, should be referred to a repair technician.

More complex problems require a schematic or service manual, an oscilloscope, and a variety of specially-made alignment and test jigs. They also require additional troubleshooting and repair techniques, which are beyond the scope of this book. If you are interested in learning more on this subject, see Appendix B for a selected list of general and specific electronic troubleshooting and repair guides.

Many manufacturers will sell you a copy of the schematic or service manual, but be prepared to spend up to $30 for it. The wait can take up to eight weeks, so order the manual *before* your deck breaks down. We provide the names and addresses of most VCR manufacturers in Appendix A. If you don't see the manufacturer of your VCR listed, or the manufacturer has moved, refer to the owner's manual or warranty card. It should list the main address of the manufacturer, plus local and regional repair centers.

Most VCR manufacturers will not sell the test jigs and alignment tools directly to consumers. In fact, many VCR manufacturers won't even sell the tools to independent service centers! Other than routine cleaning and checkout, along with replacement of some mechanical parts (most or all of which you can do yourself), all but "authorized" service centers are directed to send defective decks back to the manufacturer for repair.

TROUBLESHOOTING VCR MALFUNCTIONS

One general exception to this rule is defects in the main printed circuit board (PCB), as well as the modular sections for rf modulator, tuner, and clock/timer. These problems can often be handled in the repair shop, but they almost always entail completely exchanging the old board for a new one. In the service trade, a technician who repairs electronic products simply by exchanging a PCB is euphemistically called a "board swapper." Technicians don't much like it (the implications in the phrase as well as the inability to get parts), but individual components and ICs are not always available from the manufacturer, and they must take what they can get.

If you suspect a problem with the PCB in your deck, you might be able to get a replacement directly from the manufacturer and change the board yourself. This could save you $75 or more in labor costs. VCR makers sell the boards on an exchange basis. You send in the old, defective board—along with a check or money order to cover the service fee—and they send you a good, tested board in return. Although the replacement cost for a board varies, it is typically in the $75 to $125 range.

MAINTENANCE PROCEDURES AND SPECIAL ADJUSTMENTS

Most of the troubleshooting steps described in the charts that follow are standard maintenance procedures covered in previous chapters. Specifically, the procedures call for routine video and audio head cleaning, replacement or rejuvenation of belts, tires, and other rubber parts, and general cleaning and lubrication. These points are more fully discussed in Chapter 5, "General Cleaning and Preventative Maintenance," and in Chapter 6, "VCR First Aid."

Some repair procedures and adjustments require special techniques. You might want to attempt these repairs yourself, or refer to a qualified VCR technician. Be aware that incorrect repair or adjustment can worsen the damage of your VCR, so you should only attempt service when you feel competent in the procedure and have the proper tools. In addition, do not attempt the following adjustments if there is nothing wrong with your VCR—in other words, don't fix it if it's not broken. These adjustments and repair procedures are not part of routine maintenance.

Tape Guide Spindle Adjustment

In VHS decks, two tape guide spindles (or posts) draw the tape tight against the spinning video heads. These guides are mounted in a slotted track. The guides, shown in Fig. 9-1, serve two purposes:

- ✦ To route the tape around the head drum at the proper angle.
- ✦ To provide the correct tape height for the video head drum for proper recording and playback.

Although the track usually requires cleaning and relubrication every 12 to 18 months, the guides themselves seldom need adjustment. However, thermal expansion

Fig. 9-1. One of two tape guide spindles in a VHS deck.

and contraction (caused by normal operation of the VCR) can make the guides loose and out of adjustment.

When the guides are out of adjustment, the deck loses partial or complete tracking ability, and using the tracking control has no or limited effect. Newly recorded tapes made with the guides out of whack might look fine, but they might not be playable on a properly adjusted deck. What's more, subsequent adjustment of the guides in your VCR will render the previously made tapes unwatchable.

The height of the guides can be adjusted by loosening a set screw (usually at the bottom of the guide) and then turning the adjustment screw on top. The head of the adjustment screw is either a special slotted head or a hex (typically 1.5 mm). If your deck has a slotted head adjustment screw, you need to make a special tool by filing down the center of a flat-bladed screwdriver, as shown in Fig. 9-2.

In a service center, the adjustment of the tape guide spindles is made by connecting an oscilloscope to a test point within the VCR and then adjusting the height

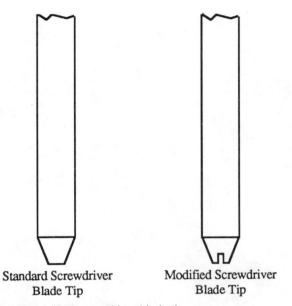

Standard Screwdriver
Blade Tip

Modified Screwdriver
Blade Tip

Fig. 9-2. Standard and modified screwdriver blade tips.

of the guides until the waveform on the scope is at its maximum peak. You can duplicate this method if you have an oscilloscope and the proper service manual for your deck (the manual should list the test point and what the waveform should look like).

Lacking these items, you can watch the picture on a good TV while playing back a commercially pre-recorded movie (or use a test tape). Center the tracking control knob on the front of the deck; then adjust the guides until all of the lines on the top and bottom of the TV screen are gone. Go slowly—a small turn makes a big difference.

Because there are two tape guides, you must adjust each one for the best picture. Write notes as you go, and if you find the picture is getting worse, turn the guide back to where it was and try the other one. Count the number of revolutions and the direction you turn the guides so you can return to where they were if necessary and start all over again.

Be extra careful when fine tuning the tape guide spindles, and then only perform the procedure when you are certain that the tape guides are out of adjustment. Don't fiddle with the guides just because you think the picture ought to "look better." If you are unsure of your abilities, take your deck to a service center and have them make the adjustment.

Control/Audio Head Adjustment

On all but the very old VCRs, the control track and audio heads are combined in one unit (VHS stereo models have both right and left channels in the head unit and might have an audio erase element as well). In service literature, this head is often referred to as the ACE head, for audio, control, and (audio) erase.

The linear distance of the head unit to the video head drum can be adjusted by turning one cone or V-shaped screw, as depicted in Fig. 9-3. By turning the screw one way or the other, the head unit is placed closer to or farther away from the video head drum, thus retarding the control track and audio signals relative to the video information. The effect adjusting this screw is tracking noise in the bottom or top of the picture (see Fig. 9-4), or loss of sound-picture synchronization.

The control/audio head does not need to be adjusted unless the heads are replaced or somehow become out of whack. The adjustment should be made using an alignment tape and by following the instructions in the service manual. But if you desire, you can adjust the head distance yourself by using a commercial pre-recorded movie as a test tape and adjusting the screw until the the audio signal sounds clear and the picture is sharp.

Fig. 9-3. The control/audio head adjustment screw (or cone).

Fig. 9-4. The effect of a maladjusted control/audio head screw is similar to mistracking and breaks up the picture into lines.

Erase Head Tape Post Adjustment

On most VCRs, either the full-erase head or the tape guide post immediately before the full-erase head is adjustable. The height of the head or guide should be checked if the full-erase head is replaced. You can always tell that the full-erase head is not completely erasing the tape if you record a new program over an old one. The video tracks will almost always be completely erased, but remnants of the control and audio tracks might remain. A breakup of the picture can mean that the old control track is not being completely erased. In addition, background sound or a loss of overall sound fidelity can mean that the audio tracks are not being erased.

The erase head or post is adjusted in a similar manner to the tape guide spindles and control/audio head, as explained above. As before, go slowly and note the results after twisting the set screw or post 1/4 or 1/8 turn.

Other Adjustments

Depending on the age and make of your VCR, there are other adjustments that you can make to repair your deck or make it run more smoothly. Most of the following adjustments are not found on VCRs made after 1982 or 1983. Check the service manual for your particular machine for the adjustment procedure:

➡ Cassette-in leaf-switch adjustment. Adjusts the switch for active contact, as shown in Fig. 9-5, when a cassette is inserted in the machine.

➡ Loading-end leaf-switch adjustment. Adjusts the switch for active contact when the loading cycle is complete and the tape guide spindles are seated against the stops.

➡ Back-tension tape adjustment. Alters the degree of back tension on the supply reel brake, as determined by the back-tension guide or lever.

➡ Supply and take-up reel torque adjustment. Adjusts the amount of torque available for REWIND, FAST-FORWARD, and PLAY operations at the supply and take-up reel or spindles.

➡ Capstan pinch roller adjustment. Adjusts the amount of pressure exerted by the capstan pinch roller.

➡ Capstan free-running speed. Adjusts the free-running speed of the capstan.

➡ Use a frequency meter or oscilloscope for precise adjustment. Some VCRs have ''timing'' marks on the belts to determine correct capstan speed. You need a neon lamp or strobe jig to make the adjustment.

➡ Video head switching-point adjustment. Adjusts the switching point between the two video heads during playback (affects SP and EP heads in a four-

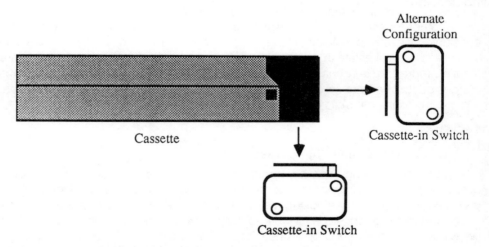

Fig. 9-5. Two possible configurations for the cassette-in switch.

TROUBLESHOOTING VCR MALFUNCTIONS

head deck, or there may be two controls for the two sets of heads in a four-head deck).

➡ Reel spindle-height adjustments. Alters the height of the supply and take-up reel spindles so that they are properly aligned. Improperly aligned reel spindles cause "scalloping" or mis-tracking. This is a very crucial adjustment and should not be made without the use of a test jig and alignment tape (most reel spindles use small spacers for proper adjustment; exchange worn-out spacers with exact-size replacements).

USING THE TROUBLESHOOTING FLOWCHARTS

The remainder of this chapter is devoted to a series of troubleshooting flowcharts that detail the possible causes and suggested solutions for a series of common and not-so-common VCR ailments. The charts are not meant to be definitive, but they should go a long way in helping you pinpoint a problem in your deck. See Chapter 8 for an explanation on how to use and interpret the charts. You are also referred to chapters 4 and 5 on the tools and supplies for VCR maintenance and repair, and how to use them.

To prevent overcomplicating the charts, they have been designed to apply specifically to ac-operated home decks. Many of the same problems, causes, and solutions will apply to portable VCRs and camcorders, but troubleshooting techniques can differ in certain situations.

You can still use the troubleshooting charts to diagnose most problems in portable or camcorder units, however, because the basic reasons behind the fault will be similar. For example, a camcorder does not have an ac cord, so you would not test its cord if you are having trouble with it. But you would test the battery or battery charger in a similar manner as you would an ac cord.

Another reason for leaning toward the home ac operated units is that they are considerably easier to maintain and repair than the portable and camcorder versions. The cramped real estate in a portable VCR or camcorder means that the parts are smaller, more fragile (generally), and harder to reach. As a matter of course, it is suggested that you not service a portable VCR or camcorder unit unless you have sufficient mechanical skills to do so.

Look to the Simple Things First

In using or looking over the charts, some of the possible causes might seem overly simplistic. The biggest mistake you can make in servicing your VCR (or anything else for that matter) is overlooking the obvious. We are reminded of the time, in our younger and more foolish days, when a color television set came in for repair. The set would turn on and you could hear sound, but there was no picture. It was an old tube set and we tore into it with a vengeance, replacing every tube that could possibly cause the problem. Alas, we replaced one tube after the other, with no results.

Then, innocently enough, we began to fiddle with the control knobs. We hit upon the brightness control, and upon turning it up, the picture magically came on! After spending well over $50 on tubes (and this was in the 1960's, so 50 bucks was a lot more money than it is now) and wasting a whole day trying to fix the set, the problem turned out to be a brightness control that was inadvertently turned down.

This was an expensive lesson, but it was a good one. It proved that if something doesn't work, odds are it's a simple cause with an equally simple solution.

List of Charts

Here are the troubleshooting charts in their order of presentation. The charts are numbered for cross-reference. Within the charts, the steps (or levels) are numbered and correspond to numbers in the text. Unless otherwise specified, all troubleshooting procedures and tests should be done with the deck turned off and unplugged (or removed from the power source).

1. VCR Does Not Turn On
2. VCR Turns On but Nothing Else
3. Cassette Will Not Load
4. Cassette Will Not Eject
5. VCR Will Not Thread Tape
6. VCR Will Not Play Tape
7. VCR Eats Tape
8. Fast-Forward or Rewind Won't Operate
9. Search Does Not Work Correctly
10. Tracking Control Has No Effect
11. VCR Does Not Respond to Some or All Front Panel Controls
12. Front Panel Indicators Not Functioning
13. Timer Does Not Operate Properly
14. Tuner Channels Don't Change on TV
15. Sound OK; Video Not OK
16. Video OK; Sound Not OK
17. Snowy Picture; Poor Audio
18. Remote Control Does Not Operate or Function Properly
19. VCR Makes Unusual Mechanical Noises During Loading and Playback
20. You Receive an Electric Shock When You Touch the VCR
21. VCR Overheats

At the end of this chapter is a set of troubleshooting guidelines for repairing miscellaneous VCR malfunctions.

TROUBLESHOOTING VCR MALFUNCTIONS

1. VCR Does Not Turn On

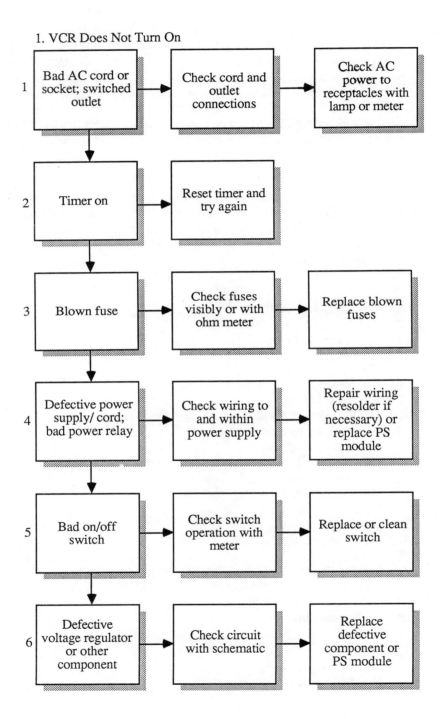

1	Bad AC cord or socket; switched outlet	Check cord and outlet connections	Check AC power to receptacles with lamp or meter
2	Timer on	Reset timer and try again	
3	Blown fuse	Check fuses visibly or with ohm meter	Replace blown fuses
4	Defective power supply/ cord; bad power relay	Check wiring to and within power supply	Repair wiring (resolder if necessary) or replace PS module
5	Bad on/off switch	Check switch operation with meter	Replace or clean switch
6	Defective voltage regulator or other component	Check circuit with schematic	Replace defective component or PS module

FLOWCHART 1—VCR DOES NOT TURN ON

The VCR is dead—nothing, even kind words, seems to revive it. Pushing the power switch has no effect, and the front panel controls, even the clock and timer, are inoperative. Here is what to look for.

Step 1-1

If the deck runs off ac power, the first logical step is to ensure that it is getting power from the ac wall source. Make sure that the cord is plugged in and that the polarized plug is inserted properly. Do not defeat the purpose of the polarized plug by filing or cutting the wide prong.

Most homes are equipped with switched outlets—you flick a wall switch to turn the outlet on and off. Sometimes the switch controls both outlet receptacles, and other times it only affects one. If the VCR is plugged into a switched outlet, make sure that the outlet is turned on.

To rule out the possibility that the problem originates in the wall outlet and not in the deck, plug a lamp into the socket and see if the lamp works. If the light glows, obviously the outlet is good. If the lamp seems dim, carefully check the voltage at the outlet with a volt-ohmmeter. It should read between 108 and 125 volts ac (117 is average). Be absolutely sure that you follow all safety precautions when testing the ac voltage or you can receive a serious shock.

Step 1-2

VCRs with timers usually shut down when in the TIMER mode (the timer starts recording at a specific time of the day). Pressing the POWER switch might do nothing, and the deck could act as if it's totally inoperative. Follow the manufacturers instructions on deactivating or resetting the timer.

Step 1-3

A blown fuse will prevent your VCR from operating. If your deck is equipped with an external fuse, remove it and visually inspect it. If you can't tell that the fuse has blown, use a meter to check continuity. Attach the leads to either side of the fuse. If the fuse is good, the meter will read 0 ohms. Anything else indicates a bad fuse.

Many fuses are located inside the deck, on or near the power supply section, as shown in Fig. 9-6. You must remove the top cover of the VCR to inspect the internal fuses. Some fuses are soldered in place but can be tested using a volt-ohmmeter. You should get a reading of 0 ohms when the test leads contact either side of the fuse. Be absolutely sure that the VCR is off and unplugged when testing fuses.

If the fuse is bad, use an exact replacement only. If the value of the fuse is 0.375 amp, do not use a 1-amp fuse.

Step 1-4

After checking the ac sources, timer, and fuse(s), inspect all power cords for

Fig. 9-6. An internal fuse, located near the transformer in the power supply section.

cracks and other signs of wear. A badly worn or damaged cord can cause a number of problems. A short will blow fuses (either in the house or in the deck); an open line will prevent the ac from reaching the deck.

Check the power cord using a meter. An open will register infinite ohms when the test leads are connected across the two prongs of the ac cord. A short will register 0 ohms when the test leads are connected to the prongs and corresponding internal wiring of the deck, as shown in Fig. 9-7. Test both prongs.

Step 1-5

The power wiring inside your deck could be faulty. Open the VCR and check the connections leading to the power switch. Check for shorts and open circuits using the meter. Normally, only one side of the incoming ac will be connected to the switch. The other side should connect directly to the power transformer.

If the deck has a relay, check it to make sure that it is getting power. The relay is usually located near the power transformer; check the heavy wires (the two smaller wires lead to the power switch and carry low voltage only). Plug the VCR in and set the meter to ac volts. Connect one test lead of the volt meter to ground and the other to the incoming wire going to the relay. There should be power. Now, attach the lead to the other power terminal on the relay. With the relay off, there should be no power; turn the VCR power switch on and the meter should register approximately 117 volts ac.

If readings are erroneous, test the continuity of both the coil and the relay terminals. The coil should remain continuous, but the contact terminals should be shorted when on and open when off (see Fig. 9-8).

If the wiring to the switch (or relay) tests OK, try the wiring leading that goes from the switch to the inputs of the power transformer (called the *primary*). Also test the wiring from the output of the transformer (called the *secondary*) to the power supply section or main printed circuit board (PCB). A typical schematic diagram of the power supply circuits for a typical VCR is shown in Fig. 9-9. Note that there are many different voltages generated by the power supply.

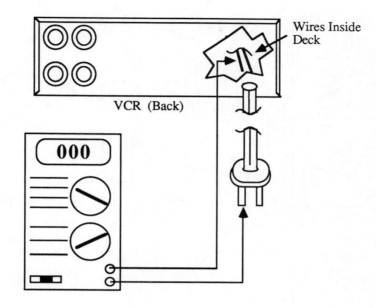

Wires Inside Deck

VCR (Back)

000

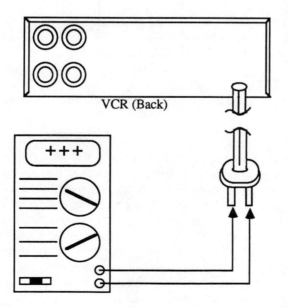

VCR (Back)

+++

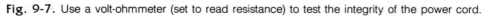

Fig. 9-7. Use a volt-ohmmeter (set to read resistance) to test the integrity of the power cord.

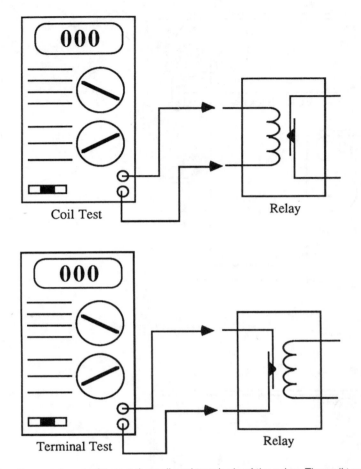

Coil Test — Relay

Terminal Test — Relay

Fig. 9-8. A meter can be used to test the coil and terminals of the relay. The coil test can be performed with power on, but be sure to use the proper meter settings and range.

To test whether the transformer is delivering power, plug in the machine (if it is safe) and turn it on. With the meter set to read ac volts, in a range of no less than 25 or 50 volts, connect the test leads to the *outputs* (secondary) of the transformer. Without a schematic, it might be difficult to make sense of the reading, but generally you should get a reading of 6 to 24 volts when connecting the leads to any of the wires.

The power transformers used in most VCRs have more than one tap-off point, so applying the leads at various tap-off terminals yields a variety of voltage levels. You can be fairly sure the transformer is working properly if you get some readings when the meter is connected to most of the secondary wires. Be sure to avoid the primary wires, as they carry the full 117 Vac from the wall outlet.

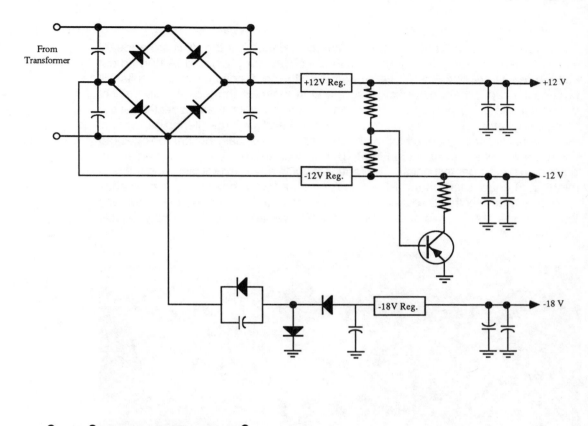

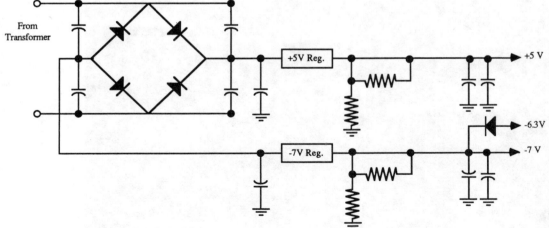

Fig. 9-9. A schematic of a representative VCR power supply. Note that the supply provides several voltages.

TROUBLESHOOTING VCR MALFUNCTIONS

Step 1-6

If you get a reading in the transformer secondaries but the VCR still appears inactive, there is probably a problem with the power supply board. A burned-out voltage regulator will cause partial or complete loss of function; similar results will happen if a capacitor or other power supply component is shorted. Visually inspect the power supply section and look for charred components. Look especially at the voltage regulators (they look like power transistors) and the large filtering capacitors.

A defective voltage regulator might be hard to spot visually but can be checked using a volt meter, as shown in Fig. 9-10. For best results, refer to a schematic to determine the proper test points. The regulator could have several pins, and the pins might not be identified on the PCB. Black charring or an oozing, dark substance around a large electrolytic capacitor indicates trouble. Replace the capacitor with the same value (both in microfarads and voltage) or replace the power supply module as a whole.

Fig. 9-10. Use the volt-ohmmeter to carefully test the operation of the voltage regulator (attached to opposite side of the board).

2. VCR Turns On But Nothing Else

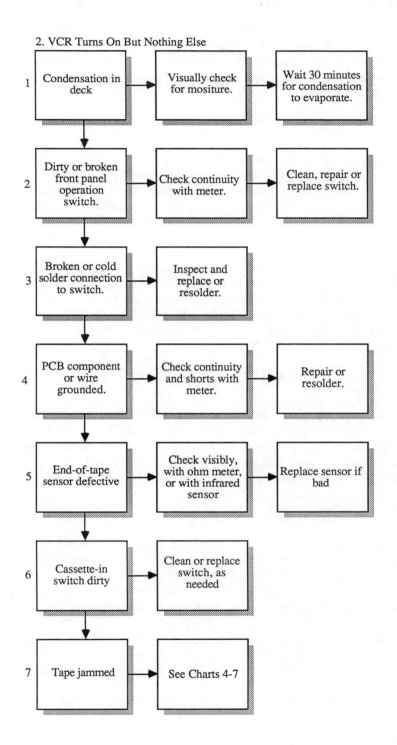

1. Condensation in deck → Visually check for mositure. → Wait 30 minutes for condensation to evaporate.

2. Dirty or broken front panel operation switch. → Check continuity with meter. → Clean, repair or replace switch.

3. Broken or cold solder connection to switch. → Inspect and replace or resolder.

4. PCB component or wire grounded. → Check continuity and shorts with meter. → Repair or resolder.

5. End-of-tape sensor defective → Check visibly, with ohm meter, or with infrared sensor → Replace sensor if bad

6. Cassette-in switch dirty → Clean or replace switch, as needed

7. Tape jammed → See Charts 4-7

TROUBLESHOOTING VCR MALFUNCTIONS

FLOWCHART 2—VCR TURNS ON BUT NOTHING ELSE

It is often harder to troubleshoot a VCR when the machine turns on but doesn't play; however, at least you can rule out problems in the power supply, fuses, ON/OFF switch, and the relay.

Step 2-1

Taking a VCR from the cold damp outside air to the warm inside air can cause condensation to form on the internal components. Many VCRs (including portables and camcorders) have a dew sensor that prevents the machine from operating when there is condensation present. If you suspect condensation, see if the DEW indicator is on (some decks don't have a dew indicator, but the STOP button indicator may flash). Also, visually inspect for moisture and wait 30 to 60 minutes for it to evaporate. You can speed up the drying process somewhat by using a hair blow dryer. Keep the dryer on low or no heat.

If the moisture does not evaporate, inspect the dew sensor. One is shown in Fig. 9-11. Its location in the deck varies, but it is always in the tape transport area. Wipe off the sensor with a clean cloth or clean it with pure alcohol (the alcohol will displace the water, speeding up the drying process). Check the leads connecting the sensor to the deck. If they are shorted, the VCR will assume moisture is present and will shut the machine off.

Step 2-2

You might not be able to control the operation of the deck if one or more of the front panel switches are dirty or broken. If only one switch is affected, you might be able to cycle the deck through its other operations. You can isolate that switch by finding its solder contacts or connecting wires on the switch panel PCB (the switch panel may be a part of the main PCB in some decks). Use the volt-ohmmeter to test the continuity of the switch. Pushing the switch should change the reading from 0 to infinite ohms or vice versa.

Step 2-3

If all the front panel switches are inoperative, the problem could lie in the common connecting wire going to the switches (if applicable) or to the front panel PCB. Test all the wires leading to the switch panel with the meter to make sure none are broken or loose. With the test leads connected to either side of the wire, a reading other than 0 ohms is an indication of an internal break in the wire. Inspect the solder points for signs of a cold or incomplete solder joint. Resolder the joint if necessary.

With many decks, the front panel PCB is attached to the main PCB by connectors, as shown in Fig. 9-12. Use your meter to test the continuity between the connectors. If you find a reading other than 0 ohms, carefully remove the connector and inspect it for loose wires, broken wires, and broken or dirty contacts. Resolder or replace the connector if it is damaged. Use a recommended cleaner to clean contact points.

Fig. 9-11. The dew sensor.

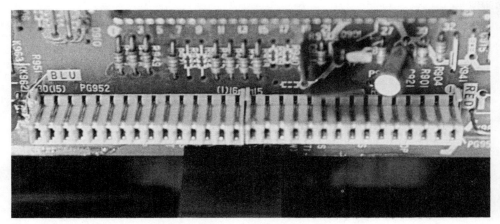

Fig. 9-12. Connector on the main printed-circuit board.

TROUBLESHOOTING VCR MALFUNCTIONS

Step 2-4

If the problem still persists, the next logical step is to examine the wires and components on the main PCB. Are there any obviously broken wires? Test with a meter to be sure. Look closely at the capacitors and resistors mounted on the board. Occasionally, you can spot a burned out component by looking for black charring on and around it.

Generally, you cannot spot a blown transistor or integrated circuit in this manner because the damage does not usually extend to the exterior of the device because of the low operating voltages of these devices. Refer to the service manual for the proper voltage levels at the pins of the transistors and ICs. Use your meter to check the voltages.

Step 2-5

In VHS decks, the end-of-tape lamp is used to indicate when the tape has reached the beginning or end. The tape normally blocks the light from the lamp to the two sensors located on either side of the transport area. But when the tape reaches the beginning or end, the light shines through the clear leader portion and strikes against a light sensitive transistor—the sensor. When the sensor receives light, playback stops and the tape is unthreaded.

If the end-of-tape lamp is out (it is usually infrared so you can't see it), the deck will still operate but the safety feature of the automatic shut-off is lost. That should be avoided to preserve the tape. Test the lamp with the infrared sensor described in chapter 4. If either of the sensors is shorted or defective or its connecting wires are shorted, the deck might assume that the end of the tape has been reached and will place itself in the auto shut-off mode.

Inspect the sensors to be sure that they are not shorted. Use your meter to measure the output voltage of each sensor (set the meter to dc volts). Block the sensor from light and the reading should be fairly low. Shine a light at the sensor and it should go up to about 5 to 12 volts. If the voltage of the sensor stays low or high no matter how much light strikes it, inspect the connecting wires for shorts. The sensor itself might be defective; replace it if so.

Beta decks do not use light to detect the end of the tape. Rather they use electromagnetic coils that detect the absence or presence of ferrous material. Two coils are used, as explained in Chapter 2, that constantly detect the ferrous surface of the tape. At each end of the tape is a small sliver of aluminum foil. When the coil touches the foil, the VCR stops playback.

You can test the coils in the same manner as the VHS lamp-sensors. Simulate a tape by placing a washer or other ferrous object on or near the coil. Watch the voltage on the meter. Now remove the object and watch for a voltage change. If the voltage doesn't change, look for shorts and inspect and test the coil. Trace the leads from the coil back to the PCB to make sure there are no breaks.

Step 2-6

All VCRs have a cassette-in switch, located on or near the cassette lift mechanism. If this switch is not activated, the VCR will assume that no tape is inserted, and none of the operator controls will function. Inspect your deck and find the switch. If the switch can't be found, refer to the service manual. When the cassette is inserted and loaded (either manually with a top-load VCR or automatically with a front-load unit), is the switch activated? Use your meter to check for continuity of the switch.

Step 2-7

If there is a tape in the deck, the tape itself could be jammed and preventing playback. See charts 5, 6, and 7 for more information on locating and servicing tape jams.

TROUBLESHOOTING VCR MALFUNCTIONS

3. Cassette Will Not Load

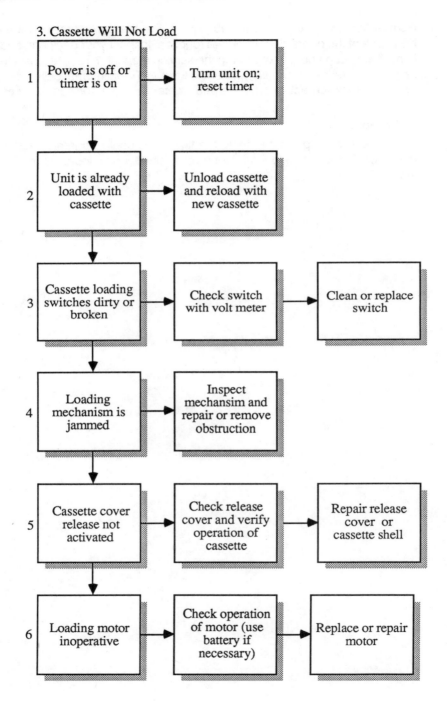

1 Power is off or timer is on → Turn unit on; reset timer

2 Unit is already loaded with cassette → Unload cassette and reload with new cassette

3 Cassette loading switches dirty or broken → Check switch with volt meter → Clean or replace switch

4 Loading mechanism is jammed → Inspect mechansim and repair or remove obstruction

5 Cassette cover release not activated → Check release cover and verify operation of cassette → Repair release cover or cassette shell

6 Loading motor inoperative → Check operation of motor (use battery if necessary) → Replace or repair motor

FLOWCHART 3—CASSETTE WILL NOT LOAD

If you find that the cassette will not load in the machine, under no circumstances should you try to force the tape. Doing so can seriously damage the VCR and will surely wreak havoc on the tape. Most of the steps in this chart assume a front-loading VCR.

Step 3-1

Is the power on? This may seem an insulting question, but remember it's the obvious things that are often overlooked. Also check to see if the deck is in the TIM-ER mode (waiting to record a show). Most decks "play dead" when in the TIMER mode; switch it to normal mode to insert the tape.

If the power is on and the timer is off, try pushing the ON/OFF button a few times to force the deck to go into EJECT mode (note: some VHS machines don't do this). The loading and threading mechanism might not have been initialized at their starting positions, and cycling the deck could put everything back to where it should be.

Step 3-2

If the VCR already has tape inside it, it certainly won't take another. Many front-load VCRs use a smoked-plastic front panel over the loading slot, making it impossible to tell if a tape is already inserted. However, most decks have a TAPE indicator that glows when a tape is inside. Don't blindly trust the indicator; it could be burned out or the sensor switch inside the deck that detects the tape could be dirty or broken.

Step 3-3

If the tape does not load and thread when you insert the cassette through the loading slot, it might not be receiving the command to do so. Figure 9-13 shows a schematic diagram of the cassette-in switch that detects the presence of the tape. Note that the loading motor will not operate unless the cassette triggers the internal leaf switch. This cassette-in switch is also used to initiate threading—no switch contact, no threading.

Make sure that the leaf switch hasn't gotten dirty, broken, or corroded. To test the switch, connect the test leads of the meter to either side of the switch terminals. The meter should alternate between 0 and infinite ohms when the switch is pushed.

Also keep an eye out for cracked or pinched wires leading to the switch, as well as poor solder joints. Test the joints and wires using the meter, as usual. Opens and shorts are indications that the wiring is bad and must be repaired.

Step 3-4

When you insert the tape, does the loading motor operate? If it does but nothing else happens, suspect the belt (if any) that connects the motor to the loading mechanism (in some machines, the capstan serves to operate loading; a belt connects the capstan to the loading mechanism).

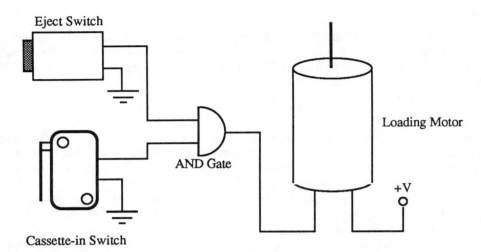

Fig. 9-13. Basic schematic of the loading motor and loading logic. The AND gate delivers power to the motor only if both the cassette-in switch and EJECT switches are on.

Belts can break or slip off the metal or plastic pulleys. Look at the belt as you insert the tape. If the motor shaft turns, but the belt doesn't budge, the belt might be too loose or slippery from oil, grease, or wear. Inspect the belt and clean it if necessary.

A number of machines use plastic or metal gears to drive the loading (or *cassette-lift*) mechanism from the motor. A gear is mounted on the motor shaft and meshes with another gear or a straight-toothed rail. Inspect the gears to be sure that none of the teeth are broken and that the gear teeth are meshing properly. Even if the loading mechanism looks fine, the gears can still bind against one another. Would oiling or greasing help? Apply a small dab of grease on the gear surfaces or oil to shafts and levers.

If the tape still won't load, inspect the path of the mechanism, including all gears, cams, slideways, and other parts in and around the loader. Carefully remove any objects that might have lodged in place.

Step 3-5

Cassettes have a release button that opens the tape cover. If this cover doesn't open, the deck can't grab the tape to thread it. The cover might not open for two reasons: the cassette itself is damaged, so pushing the release button has no effect, or the release catch in the VCR is broken or bent. Check the release catch (it's on the left of the loading mechanism for Beta and on the right for VHS) and visually see if the pin is engaging with the catch on the cassette, as shown in Fig. 9-14. Note that there is no release catch for VHS-C camcorders and that the catch is on the left for 8mm cassettes.

If the cassette is bad, you might be able to repair it using an empty shell of another discarded cassette. Transfer the tape from the bad shell to the good shell, as discussed in Chapter 5. If the release catch is bent, use a pair of pliers to bend it back into shape. The catch might need to be replaced if it is seriously bent or broken.

Step 3-6

To ensure that the proper drive signals are being applied to the loading motor when the cassette-in switch is depressed, connect the ohmmeter to both motor terminals (see Fig. 9-15). Select the dc volts function on the meter with a range of no less than 12 volts. With the deck plugged in and turned on, depress the EJECT button once (or insert a tape). With most machines, your meter should read a voltage (it can be as small as 3 volts or as high as 12 volts).

If no other defects can be found up to this point, the problem could lie in the loading motor itself. You can test the motor by connecting a C- or D-size battery to the motor terminals, as illustrated in Fig. 9-16. The polarity of the battery determines which direction the motor turns, so if the motor seems to labor when power is applied in one polarity, stop and try the other direction.

Fig. 9-14. The cover release pin, located on the side of the cassette-loading mechanism.

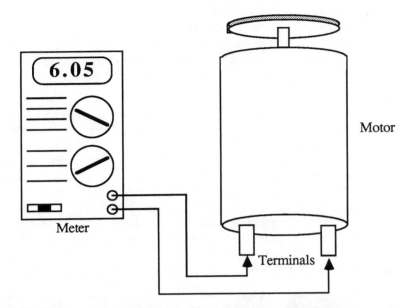

Fig. 9-15. Connect the terminals of the meter to each side of the motor.

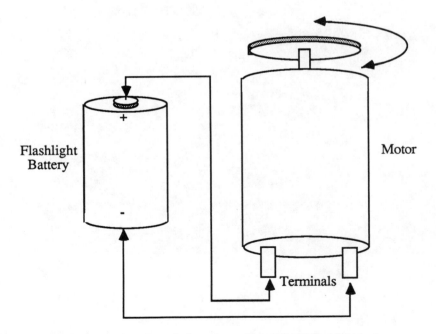

Fig. 9-16. Test the operation of small dc motors with a 1.5-volt flashlight battery.

4. Cassette Will Not Eject

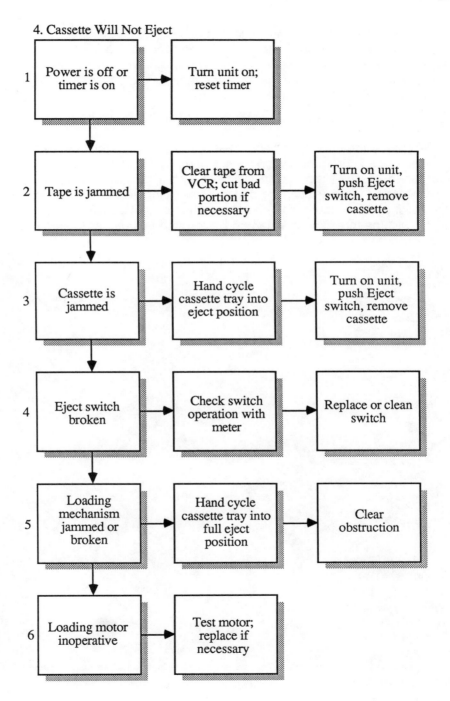

TROUBLESHOOTING VCR MALFUNCTIONS

FLOWCHART 4—CASSETTE WILL NOT EJECT

Many of the culprits responsible for the cassette not loading in a front-loading VCR are also responsible for the cassette not ejecting. Again, we remind you that under no circumstances should you try to force the tape. Doing so can seriously damage the VCR and will surely wreak havoc on the tape.

Step 4-1

Refer to the previous chart on POWER and TIMER operations. Force the eject cycle by turning the deck off and on (once should do it). Manually inspect the loading and threading mechanism to see if it is caught in mid-cycle. If it is and switching the power on and off doesn't do it, manually rotate the loading or threading mechanism to bring it back to the starting point. Figure 9-17 shows the underside of a VHS deck and how to manually roll the threading mechanism back to the unthreaded position.

Fig. 9-17. Rotate the loading motor shaft or pulley to manually engage or disengage the tape loading mechanism.

Step 4-2

Inspect the inside of the VCR for spilled tape. If it has spilled into the innards of the deck, the VCR won't be able to fully eject the tape. Manually extract the tape to eject the cassette. Hand-cycle the threading and cassette-lift mechanisms to their starting positions (see above), or turn the deck off and on. NEVER force the threading mechanism by hand by pushing on the tape-guide spindles. The same applies to the loading mechanism: never pull up on the cassette carrier.

Step 4-3

Even if no spilled tape is evident, the cassette might be stuck in the cassette-lift mechanism. Look for obstructions or blockage. Attempt to manually cycle the mechanism but do not force it. If the cassette-lift mechanism is in the up position (the cassette is not ready for threading), you might be able to simply push the cassette out with your fingers. Again, do not force it or you can cause considerable damage.

Step 4-4

The EJECT switch could be inoperable. Make sure that the switch hasn't gotten dirty, broken, or corroded. To test the switch, connect the test leads of the meter to either side of the switch terminals. The meter should alternate between 0 and infinite ohms when the switch is pushed. Keep an eye out for cracked or pinched wires leading to the switch, as well as poor solder joints. Test the joints and wires using the meter, as usual. Test the wires that lead from the external EJECT switch to the PCB or loading motor. Inspect the wires carefully and look for obvious damage.

Step 4-5

Obstructions in the cassette-lift mechanism can prevent cassette ejection. When you press the EJECT switch, does the loading motor operate? If it does but nothing happens, suspect the belt (if any) that connects the motor to the loading mechanism. Belts can break or slip off the metal or plastic pulleys. Watch the belt as you push the EJECT button. Should the motor shaft turn but not the belt, the belt could be too loose or slippery from oil or grease. Inspect the belt and clean or replace it.

Many VCRs use the same motor to operate the cassette-lift mechanism (front-loaders only) and the threading mechanism. A cam is used to cycle between the two. When a cassette is inserted, the motor first operates the cassette-lift mechanism and drops the tape downward. The cam then operates the threading mechanism, threading the tape around the heads. Inspect this cam to be sure it isn't broken. If the cam is loose, it might spin on its shaft and either not work at all or cause the deck to go into the wrong cycle at the wrong time.

Cassette-lift mechanisms that use plastic or metal gears can also be jammed. Inspect the gears to be sure that none are broken and that the gear teeth are meshing properly. Look for binding caused by poor lubrication or malalignment. Apply a small

TROUBLESHOOTING VCR MALFUNCTIONS

dab of grease on the gear surfaces or oil to shafts and levers and try re-aligning any obviously bent parts.

If the tape still won't eject, inspect the path of the mechanism, including all gears, cams, slideways, and other parts in and around the loader. Carefully remove any objects that might have lodged in place.

Step 4-6

It is unlikely, but the loading motor could be inoperative due to a burnout or broken wires. Check Chart 3 for details on how to test the loading motor for proper operation.

5. VCR Will Not Thread Tape

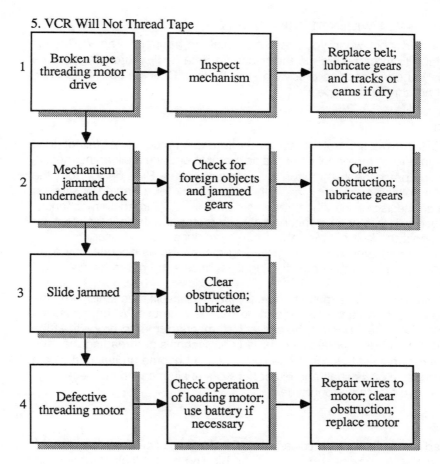

TROUBLESHOOTING VCR MALFUNCTIONS

FLOWCHART 5—VCR WILL NOT THREAD TAPE

The tape loads, but when you press PLAY nothing happens. The VCR won't engage the tape to thread it around the heads (in Beta decks, threading is done after loading; you don't need to press the PLAY button). Check these faults when the deck will not thread the tape.

Step 5-1

Some VCRs have a separate motor for the threading mechanism (all top-load decks have a threading motor only; there is no motor to operate the loading mechanism). When you press the PLAY button (or insert a tape in a Beta deck), does the motor turn and nothing happen? If this is the case, check for a broken or slipping belt. The belt connects between the threading motor and the threading mechanics. Some VCRs use rollers to convey motion between motor and threader, so look for these, too. Inspect the belt or roller and clean or replace, as necessary.

On many of the newer VCRs, the threading mechanism is operated by a set of gears. These can bind if they run dry. Lightly lubricate the gears as discussed in Chapters 4 and 5.

Many VCRs use the same motor to operate the cassette-lift mechanism and the threading mechanism. A cam is used to cycle between the two. When a cassette is inserted, the motor first operates the cassette-lift mechanism and drops the tape downward. The cam then operates the threading mechanism, threading the tape around the heads. Inspect this cam to be sure it hasn't become broken. If the cam is loose, it might spin on its shaft and either not work at all or cause the deck to go into the wrong cycle at the wrong time.

Step 5-2

The threading mechanism, particularly on older decks, can be quite complicated. It could be jammed from a foreign object or be binding due to lack of lubrication. To gain access to the complete mechanism, you'll probably need to remove the bottom panel. With older decks, you must pull out the entire tape transport assembly from the chassis, and this should be done with extreme care.

Inspect the entire mechanism and hand-cycle it through complete threading (do not force). Remove the obstructions, if any, or apply lubricant to areas that seem to be binding. Remember to go lightly on the grease; too much is almost as bad as none at all.

Step 5-3

The tape guide spindles that grab the tape and wrap it around the head might be jammed. In VHS machines, there are two tracks on either side of the video head drum where the guides slide. Check these tracks for obstructions. Clean with a cotton swab as shown in Fig. 9-18, and spread the lubricant to distribute it evenly.

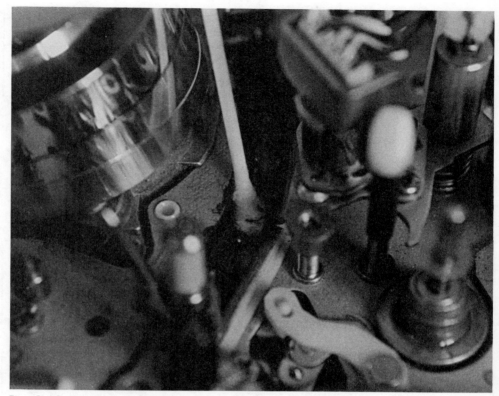

Fig. 9-18. Clean the tape guide spindle track with a cotton swab.

Beta decks use a *threading ring* to wrap the tape around the heads. Lubricate the track where the ring slides.

Step 5-4

On rare occasions, the threading motor could be defective. You can test the motor by connecting a C- or D-size battery to the motor terminals, as illustrated in Fig. 9-16, above. The polarity of the battery determines which direction the motor turns, so if the motor seems to labor when power is applied in one polarity, stop and try the other direction.

6. VCR Will Not Play Tape

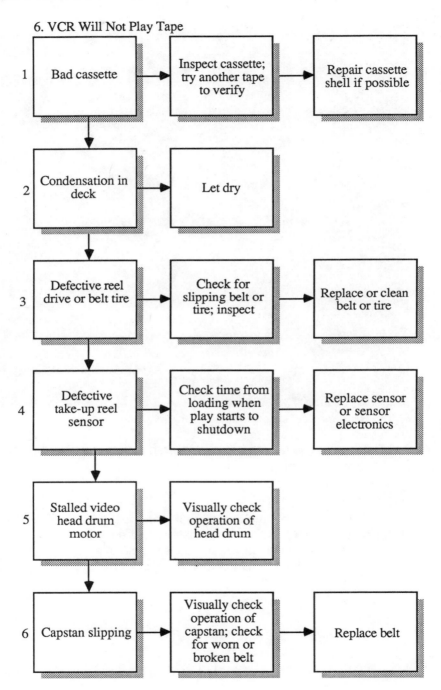

1. Bad cassette → Inspect cassette; try another tape to verify → Repair cassette shell if possible

2. Condensation in deck → Let dry

3. Defective reel drive or belt tire → Check for slipping belt or tire; inspect → Replace or clean belt or tire

4. Defective take-up reel sensor → Check time from loading when play starts to shutdown → Replace sensor or sensor electronics

5. Stalled video head drum motor → Visually check operation of head drum

6. Capstan slipping → Visually check operation of capstan; check for worn or broken belt → Replace belt

FLOWCHART 6—VCR WILL NOT PLAY TAPE

If the tape loads and threads but does not play, suspect
Be especially careful if the tape threads around the video hea
considerable tape damage could result and the heads can be

Step 6-1

Inspect the cassette to make sure it is not defective. The tape might be caught on the cover or some other component, or the internal reel locks in the shell could be preventing the tape from moving.

To test the cassette, open the cover (by pressing the release button on the side). With Beta cassettes, opening the cover automatically disengages the reel locks, so you can now turn the reels to test the free movement of the tape. With VHS cassettes, you must depress the reel lock button, located in the center on the underside of the shell. Turn the reels toward the center of the cassette. Try both reels to make sure that the tape moves freely back and forth within the cassette. Cassettes for 8mm VCRs use a similar ''release'' as VHS; VHS-C cassettes have no reel locks.

If you find the tape won't move, disassemble the shell, as described in Chapter 5, ''General Cleaning and Preventative Maintenance.''

In some cases, the tape threads and the deck attempts to play it but stops after a few seconds. This could be caused by heavy scratches on the tape that allow light from the infrared end-of-tape lamp to reach the detector. Visually inspect the tape for scratches. Try another tape or fast forward the tape until the scratch is cleared. Try again.

Step 6-2

Moisture within the deck should trigger the dew sensor, which shuts everything down to prevent damage. If the VCR is equipped with a dew light, check to see if it is on. Otherwise, visually inspect the interior of the VCR for signs of moisture. Look particularly at the polished metal surface of the video head drum. It will show condensation more readily than other parts. If dew is present, leave the VCR on for 30 minutes to one hour. Keeping the VCR powered warms up the mechanism and helps burn off the moisture.

Step 6-3

A defective reel tire or belt tire can prevent the take-up reel from spinning. This causes tape spillage, which can result in a jam. To prevent considerable damage to the deck and tape, VCRs are outfitted with a take-up reel sensor (described more fully in Chapter 2, ''How VCRs Work''). If the take-up reel spindle stops, the machine will cease playback after a 5- to 10-second delay. If the problem is mechanical—the reel isn't being driven by the belt or tire—inspect for slipping rubber and replace or clean the faulty part.

Step 6-4

The take-up reel sensor works by shining a light toward the inside rim of the take-up reel spindle. The light is reflected off strips of plastic or metal foil and directed to a sensor. If the light or sensor is malfunctioning, or the wires leading to the PCB are broken or shorted, the deck will assume the take-up reel has stopped and will place itself in auto-stop mode. Playback stops after about 5 to 10 seconds. Check the wires for breaks and shorts.

If possible, you should avoid taking the take-up reel spindle off; the height of the spindle is critical and shouldn't be altered. If you must remove the reel spindle, be sure to reassemble it with all of its parts in the correct order. Failure to replace a washer or other component will result in severe audio and video problems.

If the reel turns and all other functions look normal, inspect the sensor for dirt or oil and clean it (you must remove the take-up reel spindle for this). If necessary, replace the sensor or sensor electronics.

Step 6-5

An obstruction blocking the video head drum will cause the drum to stall out. This will prevent video playback, and most likely, will prevent the VCR from playing at all (in most VCRs, the head drum provides the timing signal to operate the drive capstan). The head drum can also be inoperative if the scanner motor or electronics are damaged.

To check for a stalled video head drum, load the tape and press PLAY. The head should immediately start to spin, even as the tape is threading. If it does not, the drum is stalled. Check for obstructions and clear them.

Step 6-6

Most all VCRs use a belt or tire to transfer rotational motion from the capstan motor to the capstan. If the belt is old or broken, the capstan will not operate properly, and playback will be impaired. If the capstan does not move at all, immediately stop the VCR. Always avoid a condition where the video heads spin but the tape does not move.

Check the capstan belt or tire and replace or clean it as necessary. While inspecting the capstan, look at the capstan motor. If it is not turning, it could be defective. Because of the delicate nature of the capstan motor, do not use a battery to test operation. Check the wiring to the motor for breaks and shorts. If you have access to the service manual, refer to it for the location of the capstan motor power terminals. Use your volt meter to check the voltage at the capstan. It will vary between 5 and 12 volts, depending on the speed of playback and the make and model of VCR.

Also check the capstan pinch roller. If the roller is old or flatted, the tape won't be pulled through the transport properly. Replace or clean the pinch roller as necessary.

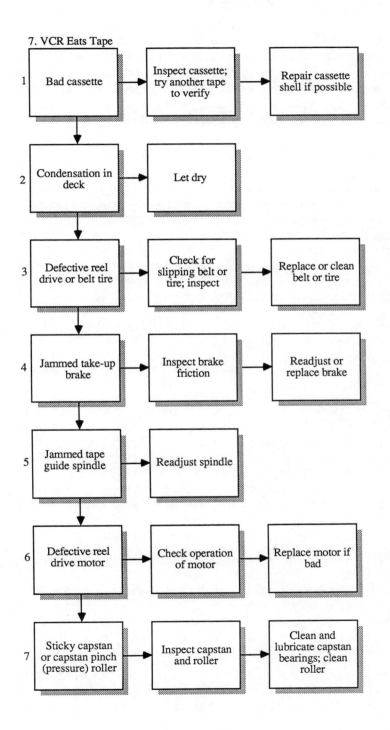

7. VCR Eats Tape

1 | **Bad cassette** → Inspect cassette; try another tape to verify → Repair cassette shell if possible

2 | **Condensation in deck** → Let dry

3 | **Defective reel drive or belt tire** → Check for slipping belt or tire; inspect → Replace or clean belt or tire

4 | **Jammed take-up brake** → Inspect brake friction → Readjust or replace brake

5 | **Jammed tape guide spindle** → Readjust spindle

6 | **Defective reel drive motor** → Check operation of motor → Replace motor if bad

7 | **Sticky capstan or capstan pinch (pressure) roller** → Inspect capstan and roller → Clean and lubricate capstan bearings; clean roller

TROUBLESHOOTING VCR MALFUNCTIONS

FLOWCHART 7—VCR EATS TAPE

Nothing is worse than a VCR with a voracious appetite. When a VCR eats a tape, the tape is usually ruined beyond repair. You can fix some tapes, but the splice can cause dirty video heads—even video head damage—during subsequent playback. The best way to avoid a VCR that eats your tapes is regular preventative maintenance. But if your VCR is having your tapes for dinner, check the possible cures below.

Step 7-1

Inspect the cassette to make sure it is not defective. The tape might be catching on the cover and is unable to come out of the shell, or the internal reel locks in the shell could be preventing the tape from moving. To test the cassette, open the cover (press the release button on the side). With Beta cassettes, opening the hatch disengages the reel locks, so you can now turn the reels to test the free movement of the tape. With VHS cassettes, you must depress the reel lock button, located in the center on the underside of the shell. Turn the reels toward the center of the cassette. Try both reels to make sure that the tape moves freely back and forth within the cassette.

If the tape won't move, it might be binding inside the shell or the locks are broken. Disassemble the shell, as described in Chapter 5, "General Cleaning and Preventative Maintenance."

Step 7-2

Moisture within the deck can cause the tape to catch on the polished surface of the head drum. Even if the deck is outfitted with a dew sensor, in some instances it will play, but if the tape stops, considerable tape stretching and damage occurs.

If the VCR is equipped with a dew light, check to see if it is on. Otherwise, visually inspect the interior of the VCR for signs of moisture. If dew is present, leave the VCR on for 30 minutes to one hour. Keeping the VCR powered warms up the mechanism and helps burn off the moisture. You can also use a hair dryer set on low or no heat.

Step 7-3

A defective idler tire will prevent the take-up reel from spinning. This causes tape spillage because the capstan continues to feed tape to the take-up reel, but the reel is not winding up the tape fast enough or not at all. Replace or clean the take-up reel idler tire. When cleaning, follow the recommended procedure outlined in Chapter 5, "General Cleaning and Preventive Maintenance."

Step 7-4

The supply and take-up spindles both have brakes (sometimes referred to as torque limiters), as shown in Fig. 9-19. In most VCRs during normal recording and playback, the supply spindle brake barely touches the reel. It is controlled by the

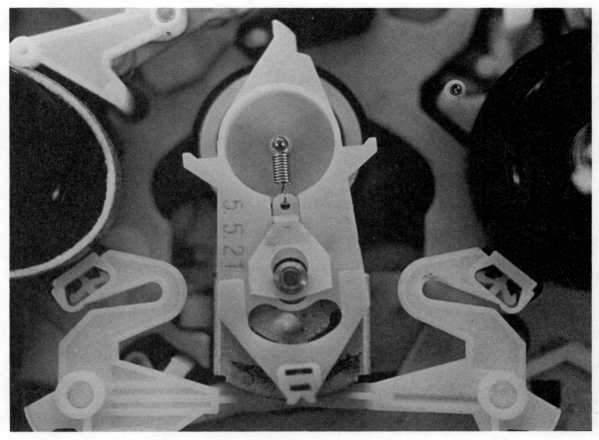

Fig. 9-19. Close-up view of the tape reels, brakes, and idler wheel pivot.

back-pressure lever. If the pressure lessens, the brake clamps to slow the supply reel; when the pressure increases, the brake releases.

Check the supply reel brake to make sure that the pad has not worn off and that there is slight pressure on the reel during operation. With the cover off, watch the tape as it passes by the back-pressure lever. It should wobble slightly.

Also during normal recording and playback, the brake on the take-up reel is set so that only clockwise rotation is possible. If it is grinding against the take-up reel, inspect the mechanism for a jam, or look for a loose or broken spring. Free the jam or fix the spring, and the brake should operate smoothly.

Step 7-5

The tape guide spindles thread the tape around the video heads and other components. If these spindles are broken or bent or out of adjustment, the tape can slip out and become tangled in the works. Inspect the guide spindles for proper operation. Use a spare tape and watch the action of the guides as you thread and unthread the tape. Keep your hand on the power switch and flick it off the moment you see the tape slip off. Manually cycle the threading mechanism back into place and eject the tape. If the guides are broken, replace them. Be careful when adjusting the spindles. They require careful adjustment, as explained earlier in this chapter.

Step 7-6

A defective reel-drive motor will have the same effect as a worn-out reel-drive idler tire. If the take-up reel doesn't turn during playback, tape spillage results. Test the operation of the motor by placing the deck in FAST-FORWARD mode. If the take-up reel doesn't move, inspect the idler tire first, then examine the motor.

Note that in most machines, however, the idler tire is driven by the capstan motor. Visually inspect the operation of the capstan motor and examine the belts from the motor to the idler tire. The drive belt is typically on the bottom of the deck, which means you must remove the bottom panel to gain proper access.

If your deck has a separate reel-drive motor, ensure that the proper drive signals are being applied to the motor by connecting a volt-ohmmeter to both motor terminals. Select the dc volts function on the meter, with a range of no less than 12 volts. With the deck plugged in and turned on, depress the FAST-FORWARD button. With most machines, your meter should read a voltage (it can be as small as 3 volts but as high as 12 volts).

If voltage is reaching the motor, it might be the motor itself. You can test the motor by connecting a C- or D-size battery to the motor terminals. The polarity of the battery determines which direction the motor turns, so if the motor seems to labor when power is applied in one polarity, stop and try the other direction.

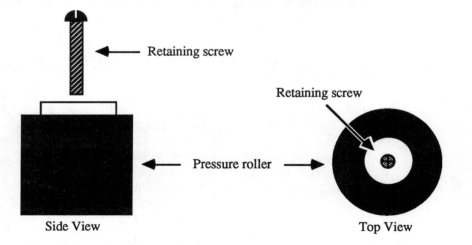

Fig. 9-20. Assembly detail of the pressure roller.

Fig. 9-21. Lubricate the capstan shaft bearings with light machine oil.

Step 7-7

A sticky capstan or pinch (pressure) roller can also cause tape spillage. With the top of the VCR off, load a tape and play it. Watch the capstan carefully and look for irregularities. Usually, problems with the capstan will show up as video and sound problems, so if you notice these, suspect the capstan or roller.

Clean the capstan and pinch roller. The pinch roller might need to be replaced or rejuvenated. See Chapter 5, "General Cleaning and Maintenance," for information on cleaning and rejuvenating capstan pinch rollers. Removing the roller for replacement or rejuvenation can be difficult because it is mounted on the support stem in different ways. Some VCRs use a screw, as depicted in Fig. 9-20, to hold the pressure roller in place. Others use an E-clip and a few use friction fit. Analyze the mounting technique before tugging on the roller.

Lubricate the capstan bearings by applying oil to the sleeve bearings, as depicted in Fig. 9-21. An oiler with a syringe applicator allows you to squirt the lubricant almost directly into the bearings. Use only a *small* amount of oil.

8. Fast-Forward Or Rewind Won't Operate

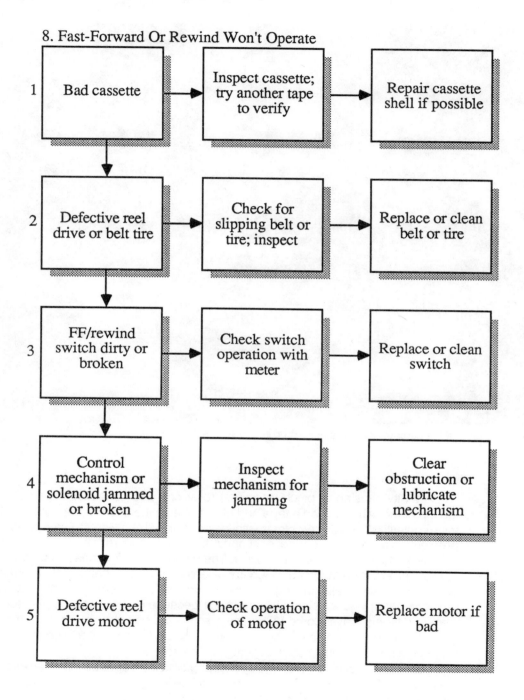

1 Bad cassette	Inspect cassette; try another tape to verify	Repair cassette shell if possible
2 Defective reel drive or belt tire	Check for slipping belt or tire; inspect	Replace or clean belt or tire
3 FF/rewind switch dirty or broken	Check switch operation with meter	Replace or clean switch
4 Control mechanism or solenoid jammed or broken	Inspect mechanism for jamming	Clear obstruction or lubricate mechanism
5 Defective reel drive motor	Check operation of motor	Replace motor if bad

FLOWCHART 8—FAST-FORWARD OR REWIND WON'T OPERATE

So, the VCR plays and records, but pressing the REWIND and FAST-FORWARD buttons has no effect. The problem is almost always related to a bad cassette, a dirty or worn-out roller, or a dirty switch.

Step 8-1

Inspect the cassette to make sure it is working properly. When in doubt, try a known, good tape.

Step 8-2

A defective idler tire can prevent the take-up or feed reels from spinning. This prevents rewinding or fast-forwarding of the tape. Replace or clean the idler tire. When cleaning, follow the recommended procedure outlined in Chapter 5, "General Cleaning and Preventative Maintenance." Also inspect the outer rim of the supply and take-up reel spindles. Use alcohol to remove any build-up of rubber and grit.

Step 8-3

The switches on even a solenoid-driven deck are inherently mechanical, and they can get broken, dirty, and corroded. Test the FAST-FORWARD and REWIND buttons to make sure that they are working properly. Connect the test leads of the ohmmeter to either side of the switch terminals. The meter should read 0 ohms when the switch is in the ON position, and infinite ohms when the switch is in the OFF position. If the switch is the double-pole type—two sets of switch contacts inside instead of just one—test each set separately.

If the switch seems OK, trace the wires back to the printed circuit board or solenoid. Use your meter to look for breaks or shorts in the wire.

Step 8-4

VCRs use two methods of activating the rewind/fast-forward mechanics: mechanical levers and electronic control. If your deck is an older model, with "piano" key controls, check the mechanism for jamming and binding. Look for foreign objects that could be blocking the smooth operation of the mechanism. Clean the linkages with a soft, clean rag and lubricate the assembly. Be on the lookout for bent parts and broken or missing springs.

Electronic controls can route to a solenoid, motor, or printed circuit board (or all three, as depicted in Fig. 9-22). Check the wires and look for shorts and open circuits. Use your volt-ohmmeter in the usual way. Observe the voltage levels at the solenoids and motor(s) when you press the FAST-FORWARD and REWIND keys. If you get a voltage, you know the circuit is working and the wiring is good; it's the solenoid or motor that's broken.

Some VCRs use a solenoid to position the idler tire in REWIND or FAST-FORWARD position. If the reel motor turns but nothing happens, check the operation of the

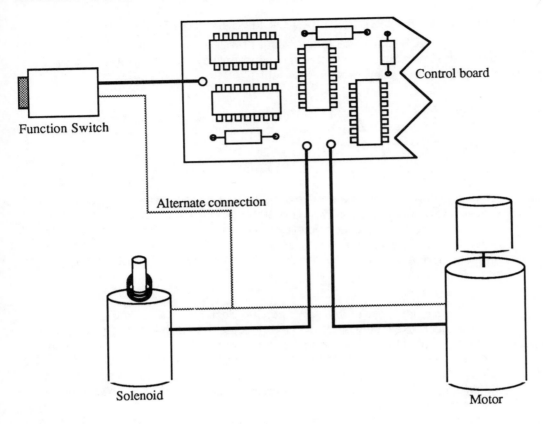

Fig. 9-22. Basic block diagram layout of a function switch and control motor/solenoid.

solenoid. Is it pulling in? If not, it might be burned out and must be replaced. If the solenoid is attempting to pull in but doesn't, look for an obstruction blocking the path of the solenoid plunger and mechanics or a broken spring or lever.

Step 8-5

In some cases, the reel motor itself will be defective. Check the voltage to the reel motor using your volt-ohmmeter. You should get a reading of between 3 and 12 volts. If there is a reading, but the shaft doesn't turn, suspect the motor. On decks that don't have a separate motor for the reel-drive (using a pick-off from the capstan motor), check the belts for dirt, glaze, or breakage.

9. Search Does Not Work Correctly

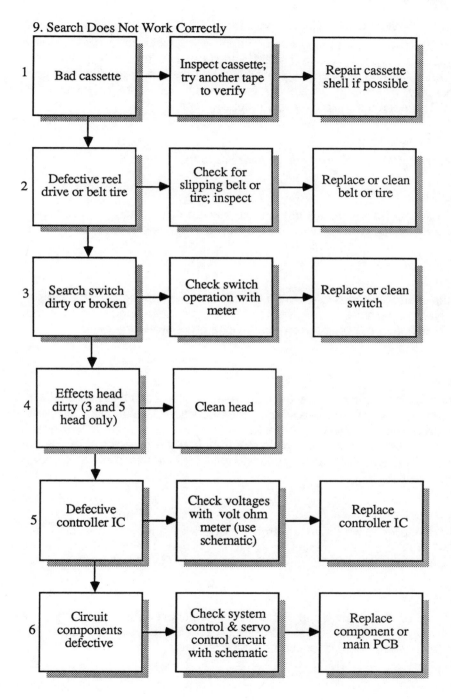

1	Bad cassette	→	Inspect cassette; try another tape to verify	→	Repair cassette shell if possible
2	Defective reel drive or belt tire	→	Check for slipping belt or tire; inspect	→	Replace or clean belt or tire
3	Search switch dirty or broken	→	Check switch operation with meter	→	Replace or clean switch
4	Effects head dirty (3 and 5 head only)	→	Clean head		
5	Defective controller IC	→	Check voltages with volt ohm meter (use schematic)	→	Replace controller IC
6	Circuit components defective	→	Check system control & servo control circuit with schematic	→	Replace component or main PCB

TROUBLESHOOTING VCR MALFUNCTIONS

FLOWCHART 9—SEARCH DOES NOT WORK CORRECTLY

Most VCRs have a fast scan or search capability that lets you quickly preview the contents of the tape. The search speed varies between models and depends on the speed of the original recording, but it is anywhere from 3 to 15 times faster than normal play. The search function can go on the fritz, however, for any number of reasons. The ones that follow are the most common.

Step 9-1

Bad cassette. As usual, try another cassette to test your theory.

Step 9-2

A defective reel-drive tire will prevent the take-up or feed reel from spinning, and therefore, the tape won't shuttle through the VCR. Replace or clean the take-up reel idler tire. Also inspect the capstan pinch roller and clean or replace it, as required.

Step 9-3

Test the search switches to ensure they are working. Connect the test leads of the ohmmeter to both sides of the switch terminals. The meter should read 0 ohms when the switch is in the ON position, and infinite ohms when the switch is in the OFF position.

Step 9-4

On three- and five-head decks, the extra head is often used for special effects. If this head is dirty, it will appear as if the search function is not working properly. Clean all heads and try again. Some decks use the full complement of four heads to achieve clear fast-scan tape searches, even though the two sets of heads are normally used for playback at various speeds. Again, clean all heads and try a known good tape.

Step 9-5

A bad controller IC will prevent searching. Check voltages with a volt-ohmmeter. You will need a service manual or schematic to perform this operation. If the voltages aren't what they are supposed to be, replace the controller IC.

Step 9-6

Additional circuit board components can be defective. As before, check the voltages with a volt-ohmmeter against the specifications in the service manual. Replace the faulty component or circuit board as required.

10. Tracking Control Has No Effect

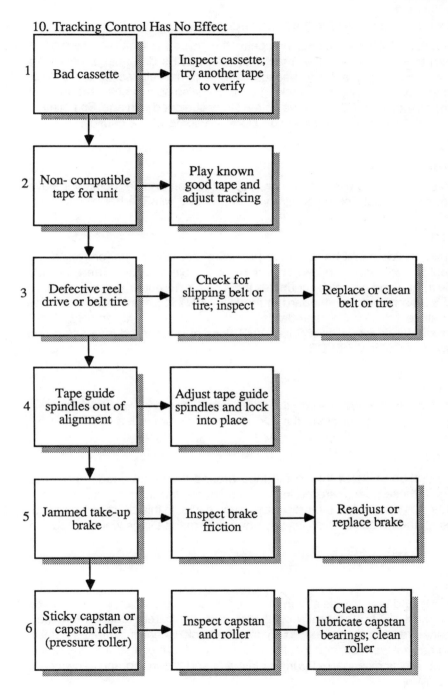

1	Bad cassette	Inspect cassette; try another tape to verify	
2	Non- compatible tape for unit	Play known good tape and adjust tracking	
3	Defective reel drive or belt tire	Check for slipping belt or tire; inspect	Replace or clean belt or tire
4	Tape guide spindles out of alignment	Adjust tape guide spindles and lock into place	
5	Jammed take-up brake	Inspect brake friction	Readjust or replace brake
6	Sticky capstan or capstan idler (pressure roller)	Inspect capstan and roller	Clean and lubricate capstan bearings; clean roller

TROUBLESHOOTING VCR MALFUNCTIONS

FLOWCHART 10—TRACKING CONTROL HAS NO EFFECT

Beta and VHS VCRs have a tracking control that helps you lock onto the picture when playing tapes made on other machines (8mm VCRs don't have a tracking control, as explained in Chapter 2, because they automatically sense the correct tracking pattern on the tape). The tracking control is also sometimes used to view tapes you made with your VCR but have gotten old or stretched with age. Sometimes, turning the tracking control has no effect. Here are some of the causes.

Step 10-1

Is the tape too bad to use? If you can, try the tape on another deck. If it plays well on another VCR but not in yours, you can proceed with the troubleshooting.

Step 10-2

Even though you might be using the proper format of cassette for your deck, the tape could have been recorded on a non-compatible machine. These include tapes made in England and other countries that use a different broadcast standard than the U.S. (the broadcast standard in the U.S. is NTSC; many other countries use incompatible broadcasting standards known as PAL, PAL-M, and SECAM).

In addition, tapes made on a Super VHS deck and recorded in SUPER mode will not play on regular VHS VCRs.

Step 10-3

A defective idler tire will prevent the take-up reel from spinning, and thus will impair playback. Replace or clean the idler tire, as recommended in Chapter 5.

Step 10-4

The tape guide spindles thread the tape around the video heads and other components. If these guides are broken or misaligned, the tracking might be off. To re-align the spindles, follow the procedure given earlier in this chapter. Be careful when adjusting the guide spindles, as you can easily make things worse if you are careless.

Step 10-5

The supply and take-up reels both have reel brakes, as depicted in Fig. 9-20, above. With most VCRs, the brake on the supply reel barely touches the reel, providing back-pressure during recording and playback. Check this brake to make sure the pad has not worn off, and that there is slight pressure on the reel. The brake on the take-up reel is disengaged during playback and recording and allows only clockwise rotation.

Step 10-6

A sticky capstan and pinch roller can also cause mis-tracking as well as profound audio and video troubles. Watch the capstan carefully and look for irregularities. Clean the capstan and pressure roller with alcohol. If the pressure roller needs rejuvenation, refer to Chart 7, above, for details. Additionally, oil the capstan bearings with a syringe applicator.

TROUBLESHOOTING VCR MALFUNCTIONS

11. VCR Does Not Respond to Some or All Front Panel Controls

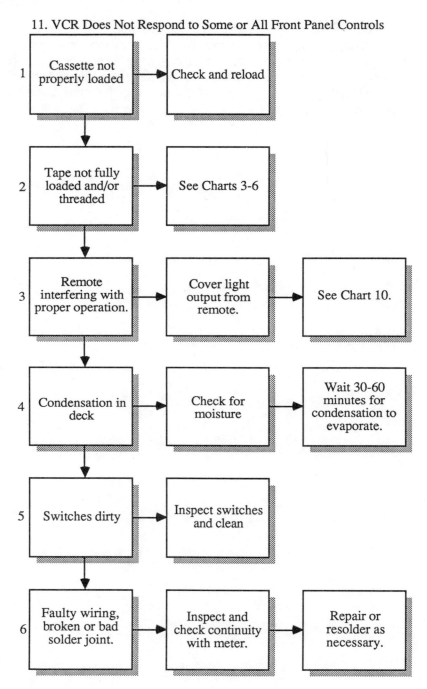

FLOWCHART 11—
VCR DOES NOT RESPOND TO SOME OR ALL FRONT PANEL CONTROLS

Failure to respond to some or all of the front panel controls is often a mechanical fault, but it can also be caused by the electronics on the main PCB. If not all of the switches are affected, the problem might lie in the switches themselves (or the functions associated with that switch). If none of the switches respond, refer to Charts 1, 2, and 6.

Step 11-1

If the tape is not properly loaded, the front panel controls might not appear to work properly. Remove the tape and reinsert it. A damaged tape can also cause the controls to behave erratically.

Step 11-2

In most VCRs, the tape must be completely loaded and threaded before the front panel controls will operate. Check Charts 3 through 6 if the tape is not loading and threading properly. Be sure to check the cassette-in switch (usually a small leaf switch, located on the inside of the cassette-lift mechanism). The switch could be dirty or broken.

Step 11-3

If your VCR has a remote control, it might be overriding the front switches. To rule out the possibility of a faulty remote control, remove its batteries, take it to another room, or—if the control is the wireless infrared type—cover up the end to block the light. If the control is the wired type, disconnect it from the deck. Refer to Chart 18 if disabling the remote frees the operation of the front panel controls on the deck (it is an indication that the remote is faulty).

Step 11-4

Taking a VCR from the cold damp outside air to the warm inside air can cause condensation to form on the internal components. Most VCRs (especially portable units and camcorders) have a dew sensor that prevents the machine from operating when condensation is present. If you suspect condensation has formed inside the machine, visually inspect for moisture and wait 30 to 60 minutes with the power on for it to evaporate. You can speed up the drying process somewhat by using a hair blow dryer, but keep the dryer on low or no heat.

Step 11-5

Given the right set of circumstances, switches can get broken, dirty, and corroded. Test each switch on the front panel to make sure all work properly. Connect the test leads of the ohmmeter to either side of the switch terminals. As you depress the switch, the meter should alternately read 0 ohms or infinite ohms. If not, it is an indication that the switch is faulty.

Unsealed switches can be cleaned using an electrical contact spray cleaner. Liberally squirt the cleaner inside the switch. Broken switches must be replaced.

Step 11-6

If the switches themselves check out, the problem might be caused by faulty wiring. Carefully inspect the wiring leading to and from the front panel control switches. Look for obvious breaks, kinks, and cold solder joints. Use a meter to test the continuity of all affected switches.

The wiring harness connecting the front panel PCB and the main PCB might be the flexible membrane type. The connecting wires are copper bands that are glued onto a flexible mylar base. These flexible ribbon connectors can be easily damaged if abused and are extremely difficult to repair. If you suspect a broken trace on the ribbon, double check the continuity using a volt-ohmmeter. A faulty ribbon must be replaced.

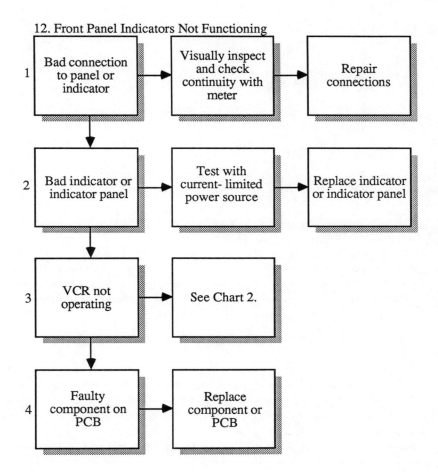

12. Front Panel Indicators Not Functioning

1 → Bad connection to panel or indicator → Visually inspect and check continuity with meter → Repair connections

2 → Bad indicator or indicator panel → Test with current-limited power source → Replace indicator or indicator panel

3 → VCR not operating → See Chart 2.

4 → Faulty component on PCB → Replace component or PCB

FLOWCHART 12—FRONT PANEL INDICATORS NOT FUNCTIONING

You put in a cassette, press PLAY, and everything works. That is, everything but the front panel indicators. Check the following points if the indicators in your VCR don't light.

Step 12-1

The indicators used in most VCRs are the LED, LCD, or fluorescent type. These have exceptionally long life spans, so it is unlikely that they have simply "burned out" like light bulb(s) do. The most probable cause is a bad wire or connection leading to the front panel or indicator.

Visually inspect all the wires leading to the front panel PCB and the indicator modules themselves. Use a volt-ohmmeter to test for continuity. With the test leads connected at the ends of each wire, the reading should be 0 ohms. A reading of

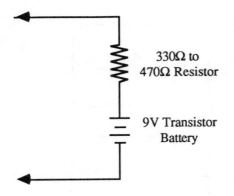

Fig. 9-23. Schematic for testing LED and LCD indicators.

infinite ohms indicates that a wire is broken or that one of the solder joints is bad. Replace the wire or resolder as necessary.

The wiring harness connecting the indicators and the front panel or main PCB might be the flexible membrane type. The connecting wires are actually copper bands glued onto a flexible base. These flexible ribbon connectors can easily be damaged and are difficult to repair. If you suspect a broken trace on the ribbon, double check the continuity using a volt-ohmmeter. Replace if bad.

Step 12-2

Rarely will all the LEDs and number segments go out all at once. The more common occurrence is that only one indicator lamp or one segment in one numeral will fail. Check first to make sure that the wiring is not at fault. Use the volt-ohmmeter as before. You can also test LED and LCD indicators and segments using the current-limited power source illustrated in Fig. 9-23. Don't forget the resistor; connecting the battery directly to the connections on the panel *will* burn out the segments.

If one or more indicators or segments are indeed bad, the entire module must be replaced. You cannot repair or replace individual indicators or segments within the module.

Step 12-3

A blank indicator panel might also be due to a fault on the deck. Be sure that the VCR responds properly to all the front panel controls. If it does not, refer to Chart 11 for more information.

Step 12-4

If all checks out so far, the fault might be a bad component on the front panel or main PCB. If the VCR works fine in every other regard but the indicator panel is blank or scrambled, the problem might be faulty driver ICs or other components on the front panel or main PCB. Should this be the case, the components or the PCB must be replaced.

13. Timer Does Not Operate Correctly

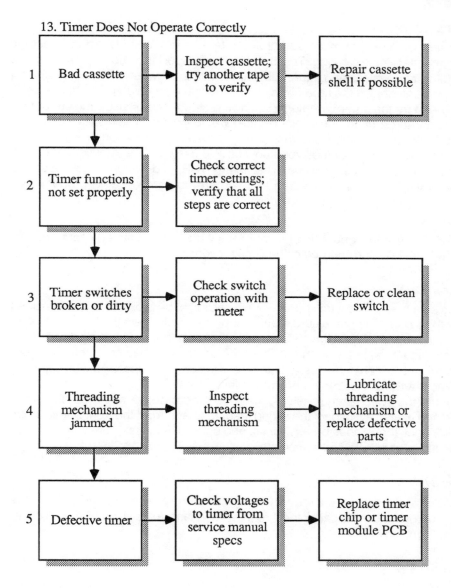

FLOWCHART 13—TIMER DOES NOT OPERATE PROPERLY

Almost all VCRs, except the very first models, incorporate a timer function where you can command the deck to record at a specific time of day (and often a specific channel). The timer feature is controlled either by the microprocessor or by discrete logic components. In most cases, programming problems are caused by mechanical defects and can be readily repaired.

TROUBLESHOOTING VCR MALFUNCTIONS

Step 13-1

As always, check to make sure that the tape is properly loaded and that the loading mechanism is working. Test the deck for proper operation by pressing the PLAY button and watch the picture (obviously, use a previously recorded tape for this, not a blank one). If the tape plays, you know that everything is fine to this point.

You can try the timer by entering start and stop times of just a minute or two. Set the timer to go off in a few minutes and wait patiently. Watch the clock on the front of the VCR and note if the machine is actuated at the appropriate time.

Of course, all this assumes you know how to use the timer function. The timer is one of the most difficult functions to use in any VCR, and it might require a careful reading of the manual to make sure that you are doing everything properly.

Step 13-2

VCRs sometimes ignore timer instructions if the timer is already set to record. Turn the timer off, and then enter new instructions.

Step 13-3

The switch(es) that control the timer functions might be broken, dirty, or corroded. Test all the switches to make sure they work properly. Connect the test leads of the ohmmeter to either side of the switch terminals. When the switch is depressed, the meter should alternate between 0 and infinite ohms. If it does not, there is a good chance that the switch under test is defective. If the switch is unsealed, you can clean its contacts with an electronic contact cleaning spray.

Should the switches themselves check out, the problem might be caused by faulty wiring. Carefully inspect the wiring leading to and from the front panel control switches. Look for obvious breaks, kinks, and cold solder joints. Use a meter to test the continuity of all affected switches.

Step 13-4

If you set the timer, go away, and on your return you find the VCR did not record the program, double-check the threading mechanism. See Chart 5 if you suspect a threading problem.

Step 13-5

Finally, if the switches and wiring check out, it's safe to assume the problem lies further on in the timer module. If the clock to the VCR is also not working, you can be fairly sure that the timer is either not functioning properly or is not connected to the main PCB or power supply of the deck. Trace the wiring to be sure that the timer is wired to the rest of the VCR. If the wiring seems OK, the timer module should be replaced.

14. Tuner Channels Don't Change On TV

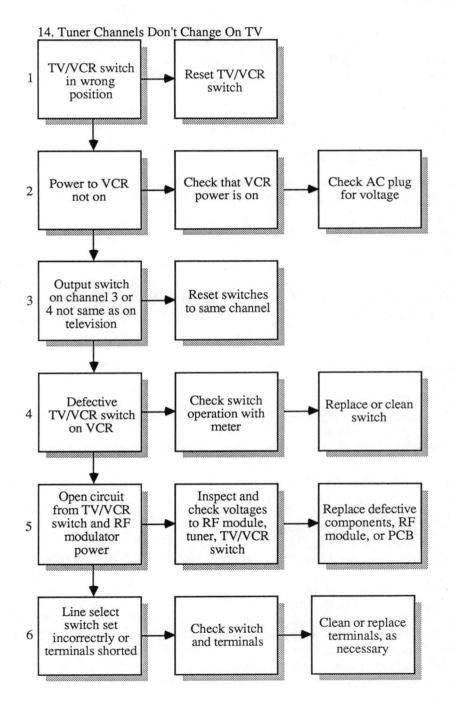

1 — TV/VCR switch in wrong position → Reset TV/VCR switch

2 — Power to VCR not on → Check that VCR power is on → Check AC plug for voltage

3 — Output switch on channel 3 or 4 not same as on television → Reset switches to same channel

4 — Defective TV/VCR switch on VCR → Check switch operation with meter → Replace or clean switch

5 — Open circuit from TV/VCR switch and RF modulator power → Inspect and check voltages to RF module, tuner, TV/VCR switch → Replace defective components, RF module, or PCB

6 — Line select switch set incorrectrly or terminals shorted → Check switch and terminals → Clean or replace terminals, as necessary

TROUBLESHOOTING VCR MALFUNCTIONS

FLOWCHART 14—TUNER CHANNELS DON'T CHANGE ON TV

You are using your VCR as a television tuner. But when you change the channels on the VCR, the channels on the TV don't change. Or all you see is static. Refer to the points below for recommended remedies.

Step 14-1

Check the position of the TV/VCR switch. When in TV position, the output of the VCR's tuner is connected to the TV. In VCR position, the tuner is disconnected (during playback) and the contents of the tape is shown.

Step 14-2

The VCR must be on for the VCR tuner to work. Be sure there is power to the deck.

Step 14-3

Many VCRs let you change the output channel between channels 3 and 4 (some even let you choose other channels, such as 5 and 6). Be sure the position of the switch agrees with the channel you are tuned to on the TV. In most instances, the switch and TV channel are set to channel 3.

Step 14-4

The TV/VCR switch can be bad. Check the switch with a volt-ohmmeter.

Step 14-5

The rf modulator or tuner sections of the VCR can be defective. First, isolate the problem by connecting the TV to the antenna or cable. Play a tape to make sure that the TV is receiving signals from the VCR. If these items check out, inspect the rf terminals on the TV for damage, shorts, and open circuits. Defective components must be replaced; the tuner and rf modulator sections are usually replaced as a whole.

Step 14-6

Most all VCRs have two sets of inputs: the RF inputs and the VIDEO and AUDIO IN inputs. The deck can handle signals presented at only one of these inputs, not both. The deck will either have a LINE SELECT switch or will have automatic line switching.

With LINE SELECT, you push the button to choose between RF input or AUDIO/VIDEO input. Check to make sure that the switch is set to the proper position, depending on how you have your VCR wired up. On machines with auto line switching, the deck will automatically choose the AUDIO/VIDEO inputs if a cable is connected to them. If you don't want to use these inputs, disconnect the cables.

If the AUDIO/VIDEO terminals on the VCR are shorted, the VCR will disengage the rf input and show only snow or a strange picture. Check the terminals to determine if shorting has occurred.

15. Sound OK; Video Not OK

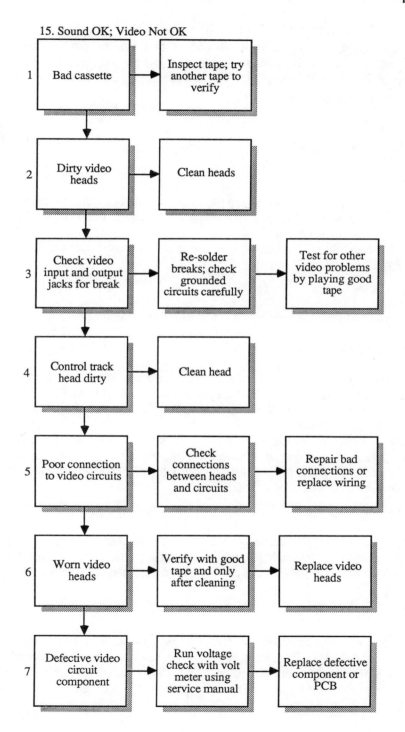

1. Bad cassette → Inspect tape; try another tape to verify

2. Dirty video heads → Clean heads

3. Check video input and output jacks for break → Re-solder breaks; check grounded circuits carefully → Test for other video problems by playing good tape

4. Control track head dirty → Clean head

5. Poor connection to video circuits → Check connections between heads and circuits → Repair bad connections or replace wiring

6. Worn video heads → Verify with good tape and only after cleaning → Replace video heads

7. Defective video circuit component → Run voltage check with volt meter using service manual → Replace defective component or PCB

TROUBLESHOOTING VCR MALFUNCTIONS

FLOWCHART 15—SOUND OK; VIDEO NOT OK

The sound plays back, but the video is either non-existent, washed-out, or snowy. Bad video is usually caused by dirty video heads, but you should also check the cable connection between the TV and VCR. Other circumstances can also contribute to poor video and are explained below.

Step 15-1

Check to see if the cassette is bad. Try a known good tape.

Step 15-2

Next to a bad tape, dirty video heads are the leading cause of video problems. Follow the procedure given in Chapter 5, "General Cleaning and Preventative Maintenance," on how to correctly clean video heads. Cleaning video heads improperly can permanently ruin them, so be sure you know what you are doing before undertaking the task.

Step 15-3

Check the cables stretching between the VCR and TV. If you are using the rf terminals on the deck and TV, be sure the coaxial cable is not crimped and that the F-fittings are on tight. Inspect the center conductor on both ends; it should be bright and shiny.

Check the cable with your meter. There should be no shorts or open circuits. Replace the cable or remake the F-fittings if they are bad (see Appendix E) or if the center conductor is dirty or broken. If you are using the direct VIDEO output of the VCR, check the cable in the same manner as above.

Dirty or broken terminals on the VCR or TV can also cause video problems. Enough of the signal might be getting through to pass the audio portion, but the connection isn't good enough for the picture. Clean the terminals using alcohol and a cotton swab and inspect them for damage.

Step 15-4

A dirty control track head can cause the video to look garbled. Clean the head and try again. In addition, try adjusting the tracking control on the front panel of the VCR to clear the lines in the screen. If the tracking control has little or no effect, refer to Chart 10.

Step 15-5

A broken or shorted wire between the video heads and the video circuitry can cause weak or no video output. The wires are usually small, so test them carefully. In many VCRs, a video amplifier is housed on or near the head drum. Check the wires from the heads to the amp and from the amp to the video processing circuitry elsewhere on the VCR. Look for bad connections (clean the connectors and make sure they are tight), broken wires, and shorts. Re-make any bad solder joints.

Step 15-6

Worn video heads can cause an overall washed-out look. Because the only cure for worn video heads is to replace them, make sure this is the problem by testing many tapes and troubleshooting all other components.

Step 15-7

A defective component in the video preamplifier or video processing circuits can produce weak or no video. To determine if a component is bad, run a voltage check with a voltmeter. Use a service manual or schematic to accurately trace the circuit. Defective parts can be replaced; otherwise, exchange the entire printed circuit board.

TROUBLESHOOTING VCR MALFUNCTIONS

16. Video OK -- Sound Not OK

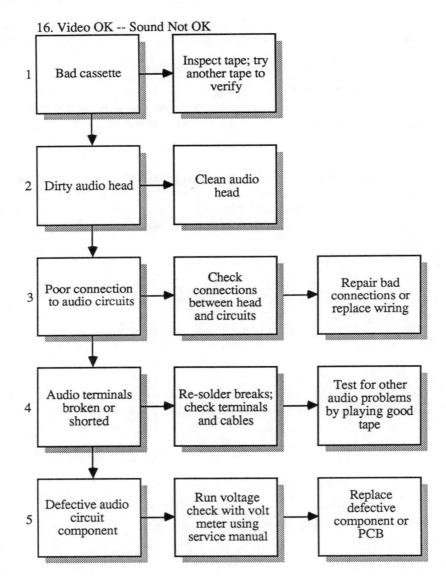

FLOWCHART 16—VIDEO OK; SOUND NOT OK

The VCR does everything it should, with one important exception: there is no audio. As you will see in this section, most audio dysfunctions are caused by dirty heads or poor cabling.

Step 16-1

A stretched or damaged tape can cause the sound to waver in and out or flutter annoyingly. In some extreme causes, the tape is so damaged that no audio can be heard. Check the operation of the VCR by using a known, good tape. If the problem persists, then you're assured that the reason lies in the VCR and not the tape.

Step 16-2

Dirty audio heads are a major cause of audio problems. Thoroughly clean the heads as described in Chapter 5, "General Cleaning and Preventative Maintenance." Linear stereo VCRs will have a single audio head unit with many elements built into it (including the control track head and both channels of audio). Remember to clean the entire surface of the head unit.

VHS hi-fi decks use two separate rotating heads to deliver hi-fi sound. Clean these as you would the video heads. Refer to Chapter 5 on how to properly clean the rotating audio heads; they are as fragile as the video heads and can be seriously damaged if mistreated. Beta hi-fi decks use the video heads to record stereo audio. If these heads are clogged, odds are that both the picture and sound will be bad.

Step 16-3

Loose audio cables between the VCR and the TV or hi-fi is a common cause of no sound. Visually inspect the cables and test them using a meter. Replace or repair bad cables. If your VCR has a headphone jack, connect a pair of phones to it. You'll hear sound if the deck is operating properly. On decks with a headphone volume control, be sure to turn the control up so you can hear the audio.

Step 16-4

The audio input or output terminals might be broken or shorted. If the terminals or the internal wires leading to them are shorted out, you will not be able to hear sound. However, you might hear a low humming sound through the speakers or TV. This is true even if your hi-fi or TV is connected to the RF output jack of the VCR and not the direct audio terminals. If this occurs, immediately turn the deck off and unplug it. Remove the cover and visually inspect in and around the audio jacks for shorted wires or foreign metallic objects.

Use a meter to test the resistance of the jacks. With the test leads connected to the inner and outer connectors of one jack, the meter should measure at least some resistance (on some decks and terminals, the reading could be near zero ohms). A reading of zero ohms is an indication that the terminal is shorted.

TROUBLESHOOTING VCR MALFUNCTIONS

If the cause cannot be found, the audio output circuit must be serviced. With some decks, the circuit is on its own printed circuit board; in others, the audio output circuitry is contained on the main PCB.

Loose, shorted, or grounded audio cables between the VCR and the monitor is a common cause of hum and distortion. Visually inspect the cables and test them using a meter. Replace or repair bad cables. If your VCR has a headphone jack, connect a pair of headphones to it. The sound should be clear and undistorted.

Excessive hum can also be caused by poorly shielded signal cables or cables routed close to ac power cords. The hum is caused by the close proximity of the alternating current field. If hum is a problem, examine the quality of the cables. The better the cable, the better the shielding, and the greater the rejection of ac-induced hum. Route the cables as far from ac cords as possible, and keep the cable lengths as short as possible.

Dirty phono jacks on the VCR or monitor can impede the signal and cause noticeable audio distortion. Clean the jacks with an electronic contact cleaner. Inspect the cable connectors; clean them too if they look dirty.

Step 16-5

Use your meter to check the wiring between the audio head(s) and the circuits on the PCB. If the wiring passes scrutiny, refer to a service manual or schematic and use the volt-ohmmeter to check for proper voltages on the circuit boards. Some decks have provisions for altering the audio input and output levels; change these controls only when you have access to the service manual data. Replace components that appear faulty or replace the entire PCB.

17. Snowy Video; Poor Audio

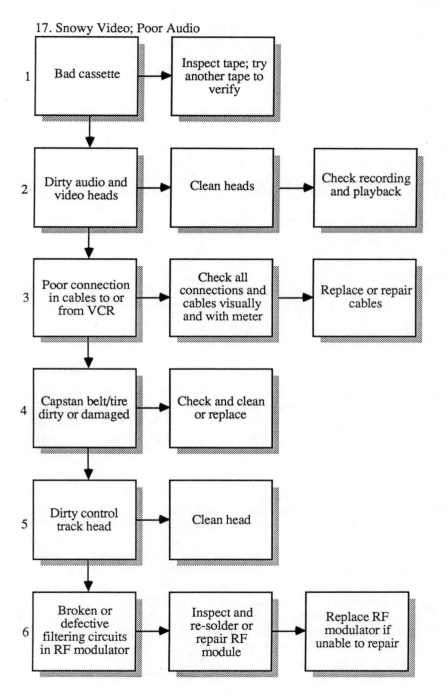

TROUBLESHOOTING VCR MALFUNCTIONS

FLOWCHART 17—SNOWY VIDEO; POOR AUDIO

When troubleshooting snowy video and scratchy audio symptons, be wary of problems in the tape, TV, and connections. It is in these areas where the problems usually reside.

Step 17-1

A stretched or damaged tape can cause serious sound and/or picture disabilities. The cassette shell might also be at fault, restricting the free movement of tape and causing excessive stress. Check the operation of the VCR by using a known, good tape (a pre-recorded movie is a convenient and functional choice). If the problem persists, then you're assured that the reason lies in the VCR, not the tape.

Step 17-2

Besides bad tapes, dirty audio and video heads are a major cause of sound and picture problems. Thoroughly clean the audio and video heads as described in Chapter 5, "General Cleaning and Preventative Maintenance." Be sure to get all of the heads. VHS hi-fi decks use two separate rotating heads to deliver hi-fi sound. Clean these as you would the video heads. Be sure to follow the directions given in Chapter 5 on how to properly clean rotating audio and video heads. They can be seriously damaged if mistreated.

Step 17-3

Unsecured audio cables between the VCR and the TV or hi-fi is a common cause of poor sound and picture. Visually inspect the cables and test them using a meter. Replace or repair bad cables. If your VCR has a headphone jack, connect a pair of headphones to it. You'll hear sound if the deck is operating properly.

Step 17-4

VCRs typically use a belt or tire to drive the capstan from the capstan motor. If the rubber of the belt or tire is worn or broken, the capstan will not operate properly and playback will be impaired. Check the capstan belt or tire and replace or clean it as necessary. While inspecting the capstan, look at the capstan motor. If it is not turning, it could be defective. Because of the delicate nature of the capstan motor, do not use a battery to test operation. Check the wiring to the motor for breaks and shorts.

A sticky capstan or pinch roller can also cause playback problems. Load a tape, play it, and watch the capstan carefully and look for irregularities. Clean the capstan and pressure roller. The pressure roller might need to be replaced or rejuvenated. See Chapter 5, "General Cleaning and Maintenance," for information on cleaning and rejuvenating capstan pinch rollers.

Removing the roller for replacement or rejuvenation can be difficult, because they are mounted on the support stem in a variety of ways. Some VCRs use a screw to hold the pressure roller in place. Others use an E-clip and a few use friction fit. Analyze the mounting technique before tugging on the roller.

Lubricate the capstan bearings by conservatively applying oil to the sleeve bearings. An oiler with a syringe-type applicator works best and helps you apply the lubricant deep into the bearings.

Step 17-5

Dirty control track heads can cause the tape to slow down and speed up or even travel through the VCR at wild speeds. Clean the control track head thoroughly and try again. Adjust the tracking control to eliminate or reduce the lines in the picture. If the tracking control can't be adjusted, refer to Chart 10.

Step 17-6

Broken or defective filtering circuits in the rf modulator can cause sound and picture problems. If your TV has direct audio and video inputs, try connecting the VCR to them. If the problem goes away, the rf modulator needs to be serviced. Inspect the modulator and re-solder or repair it, as necessary. You must replace the rf modulator if you are unable to repair it.

18. Remote Control Does Not Operate or Function Properly

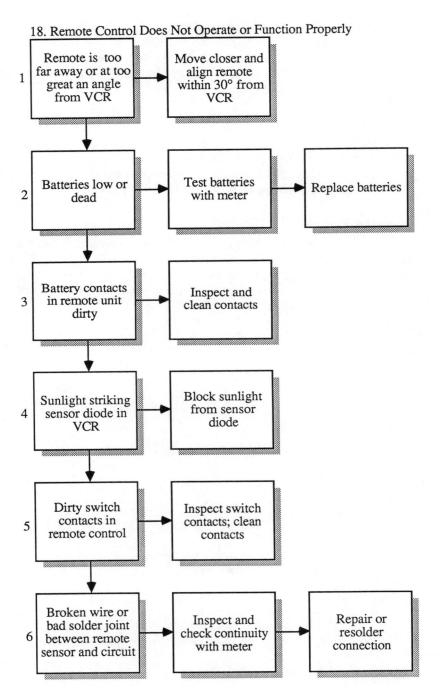

1	Remote is too far away or at too great an angle from VCR	Move closer and align remote within 30° from VCR	
2	Batteries low or dead	Test batteries with meter	Replace batteries
3	Battery contacts in remote unit dirty	Inspect and clean contacts	
4	Sunlight striking sensor diode in VCR	Block sunlight from sensor diode	
5	Dirty switch contacts in remote control	Inspect switch contacts; clean contacts	
6	Broken wire or bad solder joint between remote sensor and circuit	Inspect and check continuity with meter	Repair or resolder connection

FLOWCHART 18—REMOTE CONTROL
DOES NOT OPERATE OR FUNCTION PROPERLY

Problems with the remote control functions can be due either to a faulty remote transmitter or bad receiver circuits in the deck. Fortunately, problems with the remote control are almost always caused by the transmitter, which is not only easier to repair but cheaper to replace should it be seriously defective.

If the remote control transmitter is the infrared type, you can test for proper operation by using the light sensor described in Chapter 4, "Tools and Supplies for VCR Maintenance." If the controller is the wired type, check it as you would any switch and wire in the VCR. Use a volt-ohmmeter to check for open wires or shorts. See Step 18-6 below.

Step 18-1

If the remote control is not working properly, first make sure that you are operating the transmitter within the prescribed limits. Most transmitters don't work when operated at distances beyond 20 feet from the deck or at angles greater than 30 degrees off to either side of the receiving sensor. If you must place the deck off-angle to the direct beam of the transmitter, you can try "bouncing" the infrared light off a white card. The card is placed near the front of the VCR, as shown in Fig. 9-24.

You can also use an infrared repeater, which "re-broadcasts" the codes from the control. Position the repeater in a central location that is in-line with the controller and VCR.

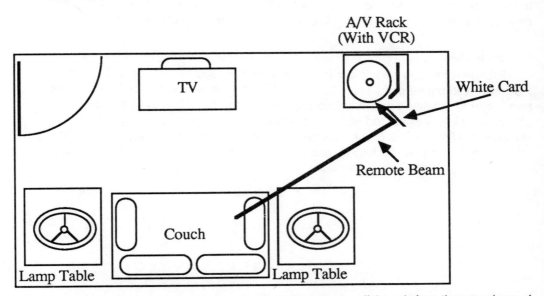

Fig. 9-24. Use a white card as a reflector when the VCR is too far off the axis from the normal use of the remote controller.

TROUBLESHOOTING VCR MALFUNCTIONS

Step 18-2

Batteries that are low or dead will obviously cause the controller to fail. Even batteries that have some kick left can cause some trouble, so even after testing them with a meter, try a fresh set to see if it makes a difference.

Step 18-3

While the batteries are out of the transmitter, inspect the battery contacts. They should be bright and shiny. If not, there might be insufficient current getting to the transmitter circuits. Clean the contacts and reinsert the batteries.

Step 18-4

Most remote controllers for VCRs work by emitting short pulses of infrared light (some of the older ones use ultrasonic sound). Sunlight striking against the deck might overload the receiving sensor, so the commands from the transmitter are not adequately received. Always shade the sensor from direct sunlight. If the deck is subjected to sunlight, the remote control function still might not operate properly even after the light has been blocked. Wait 5 to 10 minutes for the sensor and surrounding components in the VCR to cool down.

Step 18-5

Dirty switch contacts in the remote control might cause all or some of the functions to fail. Open the remote and liberally spray the switches with an electrical contact cleaner. Inspect the wiring and solder joints and make repairs as necessary.

Step 18-6

Broken wires and bad solder joints between the receiver sensor and the main PCB can also cause problems. Open the VCR only after you are certain that the transmitter is operating properly (use the light sensor described in Chapter 4 to be absolutely sure). Inspect the wires and solder joints; test the continuity with a volt-ohmmeter. With the test probes connected to either end of each wire, the reading should be 0 ohms. A reading of infinite ohms is an indication of an open circuit.

19. VCR Makes Unusual Mechanical Noises
During Loading and Playback

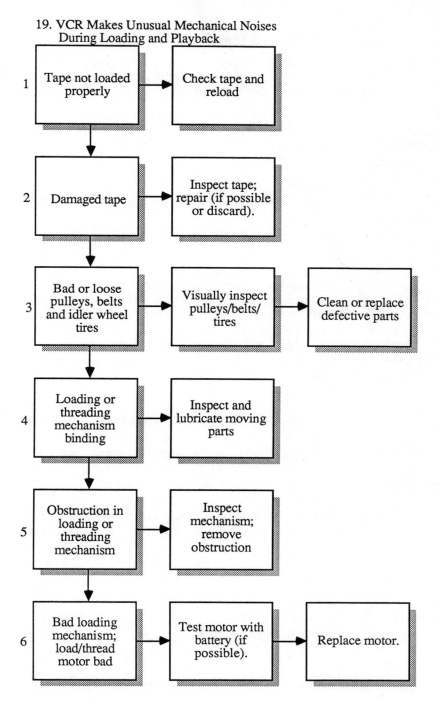

1	Tape not loaded properly	→ Check tape and reload	
2	Damaged tape	→ Inspect tape; repair (if possible or discard).	
3	Bad or loose pulleys, belts and idler wheel tires	→ Visually inspect pulleys/belts/tires	→ Clean or replace defective parts
4	Loading or threading mechanism binding	→ Inspect and lubricate moving parts	
5	Obstruction in loading or threading mechanism	→ Inspect mechanism; remove obstruction	
6	Bad loading mechanism; load/thread motor bad	→ Test motor with battery (if possible).	→ Replace motor.

TROUBLESHOOTING VCR MALFUNCTIONS

FLOWCHART 19—VCR MAKES
UNUSUAL MECHANICAL NOISES DURING LOADING AND PLAYBACK

Mechanical noises are those terrible grinding, crunching, rattling, and squeaking noises that can emanate from the VCR. A well-designed VCR is nearly silent in operation, and any noticeable noise is indicative of a malfunction inside the unit.

Step 19-1

Check that the tape is loaded properly.

Step 19-2

Is the tape damaged? Eject the tape and inspect for warpage of the shell or damage to the tape itself.

Step 19-3

Mechanical defects in belts and pulleys can cause squeaks and other unsavory sounds. Belts and pulleys are often used in the capstan drive, loading, and threading mechanisms. With the top and bottom covers of the deck off, and a tape loaded, press the PLAY button and watch the drive motor, pulley, belt, and spindle. You should be able to see any mechanical defects, if any are present.

If a belt or roller seems like it is slipping, clean it (refer to Chapter 5 for details), or replace it with a new one. Do the same when loading and unloading a tape. Watch the cassette-lift motor or drive belt and look for signs of trouble.

Step 19-4

A number of machines use plastic gears to drive the cassette-lift mechanism from a motor. A gear is mounted on the motor shaft, and meshes with another gear or a straight-toothed rail. When the motor shaft turns, the gear inches the cassette-lift mechanism along the gear or rail, thus loading or unloading the cassette. Inspect the gearing to be sure that none of the teeth are broken and that the gear teeth are meshing properly.

Even if the cassette-lift mechanism looks fine, the gears and rails can still bind against one another and cause grinding sounds. If necessary, apply a small dab of grease on the gear surfaces or oil the shafts and levers.

Step 19-5

Obstructions in the tape loading or threading mechanisms can also cause grinding or scratching sounds. Inspect the moving parts for foreign objects and remove them. Any loose parts should be retightened.

Step 19-6

The motors used to drive the loading and threading mechanisms might be bad.

A squeaky or raspy sound is caused by worn bearings. You can isolate the problem to the motor by disconnecting the drive and pressing the PLAY button. Check for any movement in the shaft that could indicate worn bearings. If the motor makes excessive noises while spinning, its a good indication that it is bad.

20. You Receive an Electric Shock When You Touch the VCR

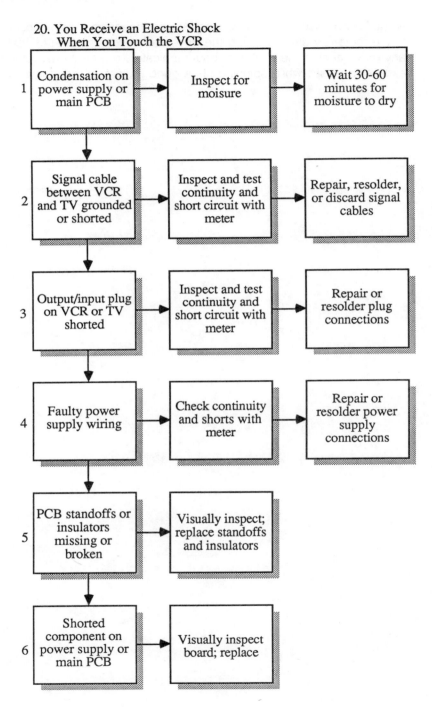

1	Condensation on power supply or main PCB	Inspect for moisure	Wait 30-60 minutes for moisture to dry
2	Signal cable between VCR and TV grounded or shorted	Inspect and test continuity and short circuit with meter	Repair, resolder, or discard signal cables
3	Output/input plug on VCR or TV shorted	Inspect and test continuity and short circuit with meter	Repair or resolder plug connections
4	Faulty power supply wiring	Check continuity and shorts with meter	Repair or resolder power supply connections
5	PCB standoffs or insulators missing or broken	Visually inspect; replace standoffs and insulators	
6	Shorted component on power supply or main PCB	Visually inspect board; replace	

FLOWCHART 20—
YOU RECEIVE AN ELECTRIC SHOCK WHEN YOU TOUCH THE VCR

The words "thrilling" and "breathtaking" are often used to describe the experience of watching a good movie on a VCR. The word "shocking" should not be in the VCR users vocabulary unless there is something wrong with the deck.

A VCR that gives you a shock—even a mild one—should not be used. Some or all of the ac power has been shorted to the metal cabinet, so when you touch the machine, you get a shock (and if the jolt is big enough, you do the "kilowatt dance," which looks a little like the jitterbug but can *really* put you in the hospital). Before watching another tape, unplug the machine, isolate the cause, and fix it.

You can test for ac leakage current with following procedures:

The ac leakage current test determines if any part of the ac line has come in contact with the metal cabinet or base. It is a safety check to prevent a potential shock hazard. It requires the use of a volt-ohmmeter.

With the VCR unplugged, short the two flat prongs on the end of the ac cord, as illustrated in Fig. 9-25. On the meter, select the 1 KΩ function and a range of no less than 1,000 KΩ. Connect one test lead from the meter to the jumper on the ac

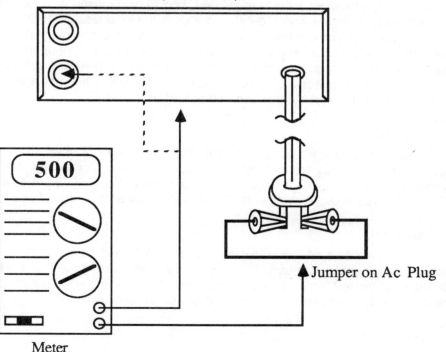

Fig. 9-25. How to test for ac leakage.

cord. Connect the other test lead to any bare (not painted) metal parts of the deck. For a typical VCR, the meter should read about 500 KΩ to 1,000 KΩ—if you get any reading at all.

A reading substantially lower than this is a good indication that the power supply has come into contact with the metal parts of the deck. If this happens, inspect the power cord as it leads into the machine, as well as the power switch, the transformer (usually bolted onto the back of the deck), and the wires leading from the transformer to the printed circuit board. If the wires are broken or are shorted to the cabinet, repair the fault before using the deck.

Not all exposed metal parts of the VCR will return a high value. Touching the center conductor of one of the input or output connectors could yield a lower resistance of 40 or 50 kilohms. This is considered normal with some machines.

An alternative method for checking leakage current is provided in Chapter 6, "VCR First Aid."

Step 20-1

Water is a poor conductor of electricity, but it's good enough to short the ac power cord or the power supply terminals against the VCR cabinet. Condensation can also form a light film of moisture that can conduct some current, and although the shock might be slight to the human body, it could damage the VCR electronics.

Inspect for moisture, and if found, let the deck dry for at least 30 to 60 minutes with the power on. You can hurry along the process by using a common hair dryer. Place the dryer in low or no heat only. If there is a lot of water, blot up the excess and spray the interior of the machine with a recommended cleaner (see Chapters 4, 5, and 6, for suggestions). The cleaner is non-water-based, and has a tendency to displace water and remove it from hard-to-reach places.

Step 20-2

The signal cable(s) between the VCR and TV might be shorted. Inspect the cabling for obvious damage and test it with a volt-ohmmeter. Connect the test leads across the center and middle conductor of each cable. You should get a reading of infinite ohms. If you get any other reading, it is an indication that the cable is shorted.

Step 20-3

Determining if the input or output terminals of the VCR is shorted is a little tougher. The resistance values can vary, but most will be within a range of about 1 KΩ to 50 KΩ when the test leads are attached across the center and outer connectors of each terminal. A reading of 0 ohms indicates a possible short.

Step 20-4

The wiring leading to and from the power supply could be faulty. Inspect the ac cord leading to the power transformer and power switch. Use your meter to check for short circuits. Examine the power transformer for obvious damage. Check the solder terminals to make sure that none are touching the cabinet. Look closely because even one small strand of wire in the ac cord can cause at least a partial short circuit.

Next, examine the wiring from the transformer to the power supply circuitry. This is sometimes located on its own PCB, other times on the main PCB. Use your meter to check for obvious short circuits. Bear in mind that at least one of the wires from the power transformer will act as the circuit ground for the VCR. You should receive a low reading when checking this wire. The other wires should yield reasonably high resistances.

Step 20-5

The power supply and main PCBs are generally insulated from the cabinet and base using plastic standoffs. If secured with metal hardware, the boards are electrically isolated from the cabinet with plastic or rubber insulators. Inspect the mounting hardware to see if any parts of the boards are touching the base or cabinet. Inspect the insulators and look for breaks and cracks. Replace any broken standoffs or insulators.

Step 20-6

Finally, a faulty component on any of the printed circuit boards in the deck can cause some leakage of current. Not all components are crucial to the operation of the deck, so even if a part shorts out, the unit might still operate—at least on a marginal level. Visually inspect the components on the boards for obvious damage, and either replace the parts (if possible) or replace the PCB.

TROUBLESHOOTING VCR MALFUNCTIONS

21. VCR Overheats

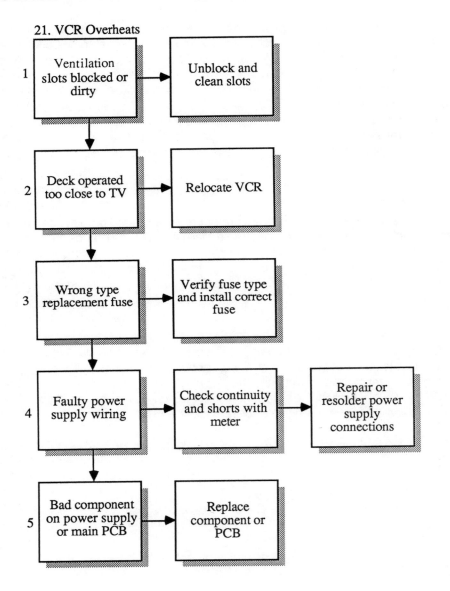

FLOWCHART 21—VCR OVERHEATS

A VCR that gets abnormally hot is not only a potential fire hazard but will exhibit erratic behavior. You can easily tell if the deck is getting warmer than normal by touching its cabinet. It should not be noticeably hot to the touch. If it is, you should investigate the problem before playing any more tapes. Continued use of the VCR might damage it beyond repair.

Step 21-1

VCRs don't consume much power, so cooling is not as critical as it is with, say, a 100-watt power amplifier. Still, most decks rely on ventilation slots for proper operation. It is important that these slots not be blocked, either by some outside object or by dust and dirt. If the slots become blocked with dust, use a brush or vacuum to clean them.

Step 21-2

A VCR operated too close to the TV might get overly warm because of the heat put out by the set. For proper operation of your VCR, as well as the other components in your video system, you should keep your TV separate as much as possible. Invest in a piece of video furniture (see Chapter 3, "The VCR Environment") to avoid stacking equipment on top of one another.

Step 21-3

Most all VCRs use external or internal fuses. If a fuse blows, it is important to replace it with the same value of fuse as the original. If a higher value is used, the fuse might not blow even though a component in the deck is shorted and drawing excessive current. That can cause fire and considerable damage to the machine.

Step 21-4

Faulty wiring in the power supply can likewise cause overheating. Check the wiring with a volt-ohmmeter against short circuits.

Step 21-5

Faulty components on the power supply board or main PCB can cause excessive overheating. You can often identify the responsible component by carefully touching each one. To test, power the deck for a while until the heat rises, then turn it off. Unplug it from the ac wall socket. Lightly touch each component. None except the voltage regulators should be hot to the touch (some ICs may be quite warm, but they should not burn your fingers). The voltage regulators, usually mounted on the power supply board, can get hot but are usually kept within a safe operating temperature by the use of aluminum heat sinks.

MISCELLANEOUS VCR DIFFICULTIES

Not every VCR malfunction can be neatly categorized. Here are a number of problems that can occur and how to remedy them.

Will Not Record

If the VCR operates well in all other respects but does not record, first check the record tab on the back of the cassette. If the tab is out, as shown in Fig. 9-26,

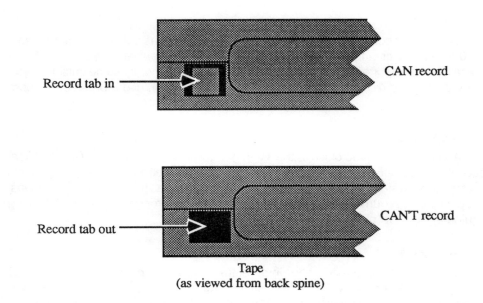

CAN record

Record tab in

Record tab out

CAN'T record

Tape
(as viewed from back spine)

Fig. 9-26. With the record tab in place, the VCR can be placed in RECORD mode. With the tab out, the VCR cannot be placed in RECORD mode.

the record-enable switch in the VCR won't let the deck erase over the tape. If you want to use the tape, place a piece of tape over the tab. Note that 8mm tapes have a slide tab. Push the tab over to prevent recording, push it back to enable recording.

Sometimes the record enable switch in the VCR becomes dirty or out of alignment. Check its operation with a meter. Visually examine the operation of the switch when you insert a tape into the deck. When the record tab is in place, the switch should be actuated (VHS and Beta). If the switch is not activated, it could be out of adjustment. Loosen the mounting screw(s) and readjust.

Color Sometimes Flashes Off and On

When you are watching a tape, does the color sometimes blank out for a moment, then come back on? Although this problem might be rooted in the anti-copy signals encoded on the tape, it can also indicate a worn or flattened idler tire or pinch roller. Inspect the idler tire and pinch roller for excessive wear, grease build-up, glaze, or cracking. Replace or rejuvenate the rubber, as explained in Chapter 5, ''General Cleaning and Maintenance.''

A flattened pinch roller, like that in Fig. 9-28, will cause the color to come and go at a fairly quick rate, but the repetition will vary depending on the playback speed of the tape (the tape will almost flicker in SP speed but slowly pulse in EP speed). Visually check the roller for a flattened end and replace it.

A maladjusted television set is another cause of flaky color reception. Be sure you are fine-tuned to the output channel of the VCR. In rare cases, the VCR itself needs adjustment. One or more internal controls determine the color output intensity and timing. Refer to the service manual for your deck for more information.

Weak Video and Audio Playback

Magnetized heads can diminish the output strength of video and audio heads. Most VCR manufacturers now recommend against demagnetizing the video heads, but the audio heads should be routinely demagnetized using a demagnetizing tool. For video heads, almost all VCR makers suggest you use a demagnetizing tool rated specifically for video work, not a regular audio deck tool. Demagnetizers for audio machinery are often too powerful for use inside VCRs.

Nevertheless, video heads do become magnetized, requiring demagnetization. Improper use of the demagnetizer tool can shatter the video heads, so be sure you are careful and go slowly. Read Chapter 5, "General Cleaning and Preventative Maintenance" for more information on using demagnetizers.

Sparkles in Playback

Most all VCRs made since the early 1980's use an electrostatic discharge element on the video heads (see Fig. 9-28). This element helps neutralize the static that can build up in the head as the tape rubs past it. If this discharge element is missing, broken, or not making sufficient contact, the video playback can be riddled with flashes of light. Recordings might not be as good as they can be, and if your deck is a hi-fi model, the sound might crackle and pop.

Examine the discharge element and make sure it is securely mounted. The tip of the element should touch (or barely touch) the center of the video head drum. If there is considerable distance between the tip of the element and drum, use a pair of pliers to bend the metal so that better contact is made.

Beta Mis-Threading

Beta decks use a novel "threading ring" to grab the tape and press it against the video and audio heads. A main roller brings the tape out of the cassette and wraps it around the head. When this roller is worn or becomes misadjusted, threading

Good Pressure Roller

Flattened Pressure Roller

Fig. 9-27. A good pressure roller versus a flattened pressure roller.

Fig. 9-28. Electrostatic discharge unit on top of head drum.

problems can occur. Check the condition of the main roller (also called the lead guidepost) and adjust or replace it as necessary.

Bent or Broken Coil Brackets

In Beta VCRs, the end-of-tape sensors are coils mounted on brackets on either end of the tape transport area. The coils work by sensing the presence of a magnetic flux, created by the coating on videotape. When the flux disappears, as it does when the aluminum strip at the ends of the tape are encountered, the VCR stops playing.

Since the coils must be in proximity to the tape to work, misalignment can sometimes cause the deck to prematurely turn off. Check the coil brackets to make sure that they are not bent or broken. If a bracket is broken, you can *temporarily* correct the problem by taping a small metal washer to the coil. Do not make this a permanent fix, because you loose the safety device of auto shut-off at the end of the tape.

Tape Counter Skips or Doesn't Work

A counter is used to show the relative position of programming on the tape. On older Beta and VHS decks, the counter is mechanical, connected to a belt driven off the take-up reel. On all newer VCRs, the counter is electronic, and usually borrows the sensor built into the take-up spindle. When the spindle turns, an infrared lamp and detector sense the rotation and increment or decrement the electronic counter.

In the mechanical system, the counter can stop working if the belt becomes worn or broken. Replace the belt and all should be fine. Almost all tape counters have a RESET button; that button is spring loaded and when pushed, resets the digits to zero. Sometimes, usually through abuse, the spring in the counter can get broken and the counter will no longer reset. You can try dismantling the counter and replacing the spring (a suitable replacement might be found at the hardware or hobby store) or you can replace the entire counter as a whole.

Failures in the electronic counters can also be accompanied with other problems. For example, if the sensor system for the take-up reel is broken, the VCR will automatically shut off after 5 to 10 seconds when it realizes that the spindle isn't moving.

If the sensor is indeed working, look to shorts or opens in the wiring leading from the sensor to the counter. Use your ohmmeter to check for trouble. Also check the RESET button by the counter. Like mechanical counters, the button used with electronic counters also has springs. If the spring is broken, the button contacts can be jammed. That causes the counter to remain on "0000," even when the tape moves.

On a few VCRs, the electronic counter is driven by a separate sensor module. The module is what's technically called a "shaft encoder," and is like the sensor employed in the take-up reel spindle. When the encoder turns, an infrared LED and sensor detect movement and direction, as illustrated in Fig. 9-29. This information is supplied to the counter electronics. The encoder is driven by the take-up reel, so check the belt or roller that connects them. Clean or replace as necessary.

Also check the sensor to make sure that it is outputting a signal. You can use a logic probe for this or a meter. Connect the encoder sensor as shown in Figs. 9-30 and 9-31, depending on whether you are using a logic probe or meter. Remember that most meters do not handle voltage changes very quickly, so manually turn the encoder shaft slowly with your fingers.

Picture Flags

Have you ever noticed an unusual bending at the top of the screen? Video technicians usually refer to this phenomenon as "flagging." It has many causes, some of which you can't correct:

➡ The tape tension is wrong. If the tape tension is too tight or too loose, the picture might flag at the top. Examine the supply reel back tension control and brake and try a new tape.

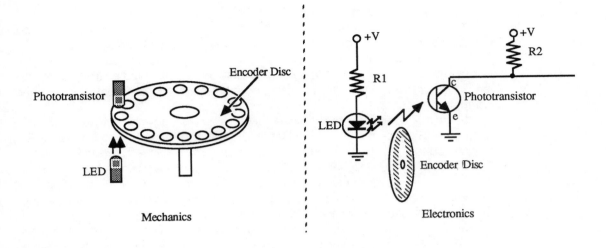

Fig. 9-29. How a shaft encoder works. The disc might have holes or slits in it, or can contain reflective strips (the LED and phototransistor are placed on the same side).

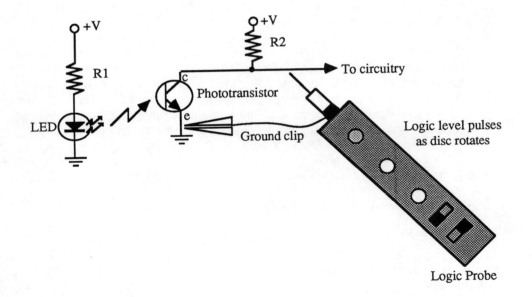

Fig. 9-30. Connect the logic probe to the output (usually collector) of the phototransistor and watch for pulses as shaft encoder rotates.

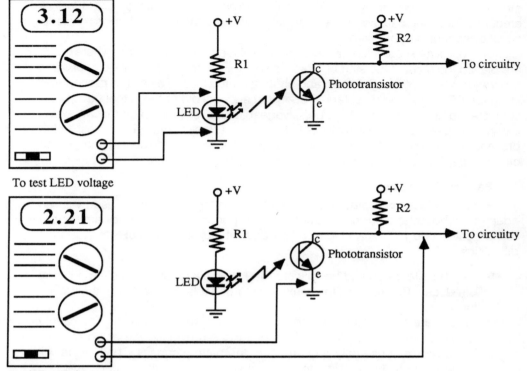

To test LED voltage

To test phototransistor output

Fig. 9-31. Connect the meter to the LED or phototransistor to test for voltage levels.

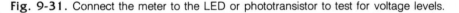

▸ The tape is recorded with an anti-copying signal. The first anti-copying techniques altered the vertical timing signals required by a VCR to lock onto the picture during recording. Unfortunately, even when you are not copying a tape, some TVs exhibit flagging due to timing instability problems. The loss of good vertical sync also upsets the horizontal synchronization that follows. This problem is most prevalent in American-made television sets manufactured before 1980. A video stabilizer that totally rebuilds the vertical synchronization signal has been known to help solve this problem.

▸ The tape has stretched. A stretched tape can exhibit flagging on the top of the screen. Also, the audio and overall picture can be bad.

Excessive Dropouts

In this instance, the term "dropout" has nothing to do with itinerant youth, but it involves a tape that is losing its magnetic coating. Loss of oxide results in a partial picture loss, which appears on the screen as a white or black "fleck." Dropout is mostly compensated for by the VCRs built-in dropout compensator, but a heavy amount of dropout, caused by relatively large areas of the tape with missing oxide coating, cannot be corrected.

Dropout is most prevalent at the beginning of tapes. This is the area where tapes are played the most. If the start of a tape is severely distorted by dropout, consider clipping out the offending portion and shortening the tape. Follow the procedures in Chapter 5 on splicing videotape. Tapes that are very old or those that have been subjected to heavy use might not be salvageable and will have to be thrown away.

Avoiding excessive dropout is not only a good aesthetic practice, but it also helps prevent the video heads from becoming clogged. The more oxide coating that comes off the tape, the more material there is to dirty the video heads.

Battery Problems

Portable VCRs and camcorders operate off batteries and when used with an adapter, off household current. Camcorder and portable VCR batteries are usually rechargeable (the nickel-cadmium, or "ni-cad," type) and are subject to certain difficulties.

➠ Ni-cad batteries should be fully charged before use, or their life-expectancy is reduced. Before using the camcorder or portable, fully recharge the battery for the recommended period of time.

➠ Ni-cads are subject to a phenomenon known as "memory effect." The battery tends to "remember" the duration of service previously expected of it, so if you regularly use only 50 percent of the battery capacity, battery life will be severely reduced. Whenever possible, "use up" the battery until it is completely dead. One way to do this is to leave the portable or camcorder on and let the battery run dead.

➠ Batteries should be regularly recharged, even if they are not used. Recharge batteries every two to four months, or they can be permanently damaged.

➠ Some camcorder battery packs use lead-acid or gelled electrolyte batteries. These, like ni-cads, are sealed against leakage, but given the right circumstances all batteries can leak. When not in use, remove the battery and store it in a cool, dry place.

If the portable VCR or camcorder is not receiving battery power, check to make sure that the battery is properly charged. You can test the voltage of a battery to see if it is working properly, but this test might not be effective unless the battery is currently in use. Insert the battery in the VCR or camcorder and turn the unit on. Use your meter to measure the voltage output of the battery. It should be roughly the same as the rated voltage.

Check the battery terminals to make sure they are not corroded or broken. Clean the terminals if they are dirty or corroded; heavy corrosion or breakage must be repaired by replacing the terminals or battery compartment module. Use your meter to check the wiring from the battery compartment to the main board on the VCR or camcorder. You should get a reading all the way to the ON-OFF switch.

10

VCR Reference Guide

This chapter contains information on over four dozen popular home and portable video cassette recorders. Included are basic features (to help you during maintenance and repair), critical maintenance points, and where appropriate, manufacturer cross-reference (similar models). They are listed in alphabetical order by brand and numerically within each brand category.

The specifications listed are from the manufacturer. These include type of remote (if any): IR for infrared or WIRED for the tethered remote; whether a timer is supplied and its limitations, for example ''Day/Event: 14/4'' means you can program up to 4 different programs over the next 14 days; and whether the VCR has high quality (HQ) circuits (discussed in Chapter 1). The notations for the critical maintenance points were drawn from repair records made on several hundred machines and reflect the common malfunctions that occur in each VCR. Your deck might not develop all of these same malfunctions, but it could indicate where extra preventative maintenance is required.

The manufacturer cross-reference logs the original maker of the deck (called an OEM, or Original Equipment Manufacturer), if any, and might indicate a ''parent'' VCR model as well as similar models sold by other companies. It is important to remember that the cross-referencing might not specifically reflect the design of the unit specified. Even though the decks might look alike on the outside, they could be considerably different on the inside. The differences can be subtle or great, but the information should help you in finding replacement parts and suitable service literature.

VCR REFERENCE GUIDE

Brand: Emerson
Model: VCR 870
Format: VHS
Type: Tabletop
General Specifications:
    Number of Video Heads: 2
    Additional Rotating Heads: None
    Remote: IR
    Linear Stereo: No
    Hi-Fi Stereo: No
    Audio Dub: No
    Timer (Day/Event): 14/4
    HQ Circuits: No
Maintenance Points:
    Worn idler tire can cause intermittent or complete cassette loading and unloading problems.

Brand: Fisher
Model: FVH-520
Format: VHS
Type: Tabletop
General Specifications:
    Number of Video Heads: 2
    Additional Rotating Heads: None
    Remote: N/A
    Linear Stereo: No
    Hi-Fi Stereo: No
    Audio Dub: No
    Timer (Day/Event): 7/1
    HQ Circuits: No
Maintenance Points:
    Spindles can become loose and cause poor or no video recording and playback (audio will remain). Adjust the spindles to regain the picture.

Brand: Fisher
Model: FVH-615
Format: VHS
Type: Tabletop
General Specifications:
    Number of Video Heads: 2
    Additional Rotating Heads: None
    Remote: N/A
    (continued)

Linear Stereo: No
Hi-Fi Stereo: No
Audio Dub: No
Timer (Day/Event): 14/4
HQ Circuits: No

Maintenance Points:

Worn drive belt can cause squeaking, tape jamming, or malfunction in rewind and fast-forward where the unit won't stop, even after STOP button is pressed. Spindles can become loose and cause poor or no video during recording and playback (audio will remain). Adjust the spindles to regain the picture.

Brand: Fisher
Model: FVH-720
Format: VHS
Type: Tabletop
General Specifications:
Number of Video Heads: 2
Additional Rotating Heads: None
Remote: IR
Linear Stereo: No
Hi-Fi Stereo: No
Audio Dub: No
Timer (Day/Event): N/A
HQ Circuits: No

Maintenance Points:

Worn drive belt can cause squeaking, tape jamming, or malfunction in rewind and fast-forward where the unit won't stop, even after STOP button is pressed.

Brand: Fisher
Model: FVH-904
Format: VHS
Type: Tabletop
General Specifications:
Number of Video Heads: 2
Additional Rotating Heads: None
Remote: IR
Linear Stereo: No
Hi-Fi Stereo: No
Audio Dub: No
Timer (Day/Event): 14/4
HQ Circuits: No

(continued)

Maintenance Points:

Worn rubber belts and rollers can cause playback and recording problems; recordings might not be stable and this could appear to be a dirty-head or servo problem.

Manufacturer Cross-Reference:

Model FVH-905 similar, includes HQ circuits.

Brand: General Electric
Model: 1VCR6004X
Format: VHS
Type: Tabletop
General Specifications:

Number of Video Heads: 2
Additional Rotating Heads: None
Remote: IR
Linear Stereo: No
Hi-Fi Stereo: No
Audio Dub: No
Timer (Day/Event): 14/4
HQ Circuits: No

Maintenance Points:

Loading mechanism can shift out of adjustment, preventing operation of REWIND and FAST-FORWARD modes. Also, cold solder joints in head amp area can cause intermittent color playback problems.

Manufacturer Cross-Reference:

GE models are made by Matsushita; Model 1VCR6010X is same but with channel presets and noise reduction; models 1VCR6012X and 1VCR6011X with two and four heads, respectively.

Brand: Hitachi
Model: VT-18A
Format: VHS
Type: Tabletop
General Specifications:

Number of Video Heads: 2
Additional Rotating Heads: None
Remote: N/A
Linear Stereo: No
Hi-Fi Stereo: No
Audio Dub: No
(continued)

Timer (Day/Event): N/A
HQ Circuits: No

Maintenance Points:
Worn idler tire causes slippage during play and record; colors can shift in and out and tape jamming and spillage can result.

Manufacturer Cross-Reference:
Hitachi VCRs are made by Hitachi.

Brand: Hitachi
Model: VT-19A
Format: VHS
Type: Tabletop
General Specifications:
Number of Video Heads: 2
Additional Rotating Heads: None
Remote: IR
Linear Stereo: No
Hi-Fi Stereo: No
Audio Dub: No
Timer (Day/Event): N/A
HQ Circuits: No

Maintenance Points:
Worn idler tire causes slippage during play and record; colors can shift in and out and tape jamming and spillage can result.

Manufacturer Cross-Reference:
Hitachi VCRs are made by Hitachi.

Brand: Hitachi
Model: VT-33A
Format: VHS
Type: Tabletop
General Specifications:
Number of Video Heads: 2
Additional Rotating Heads: None
Remote: IR
Linear Stereo: No
Hi-Fi Stereo: No
Audio Dub: No
Timer (Day/Event): N/A
HQ Circuits: No

(continued)

Maintenance Points:
Worn idler tire causes slippage during play and record; colors can shift in and out and tape jamming and spillage can result.

Manufacturer Cross-Reference:
Hitachi VCRs are made by Hitachi.

Brand: Hitachi
Model: VT-8000A
Format: VHS
Type: Tabletop
General Specifications:
Number of Video Heads: 2
Additional Rotating Heads: None
Remote: IR
Linear Stereo: No
Hi-Fi Stereo: No
Audio Dub: No
Timer (Day/Event): N/A
HQ Circuits: No

Maintenance Points:
Unit uses many rubber belts and tires. When dirty or worn, can cause tape spillage, threading malfunctions, and no video playback (audio will remain, but might be garbled).

Manufacturer Cross-Reference:
Hitachi VCRs are made by Hitachi.

Brand: Hitachi
Model: VT8500A
Format: VHS
Type: Tabletop
General Specifications:
Number of Video Heads: 2
Additional Rotating Heads: None
Remote: IR
Linear Stereo: No
Hi-Fi Stereo: No
Audio Dub: No
Timer (Day/Event): N/A
HQ Circuits: No

(continued)

Maintenance Points:

Failed relay (RL402) can cause rewind and play failure; replace or clean relay (clean and lubricate with contact cleaner).

Manufacturer Cross-Reference:

Hitachi VCRs are made by Hitachi.

Brand: JVC
Model: HR-142U
Format: VHS
Type: Tabletop
General Specifications:

Number of Video Heads: 4
Additional Rotating Heads: None
Remote: IR
Linear Stereo: No
Hi-Fi Stereo: No
Audio Dub: No
Timer (Day/Event): 14/1
HQ Circuits: No

Maintenance Points:

Tape sensor lamp can burn out, causing non-operation.

Brand: Lloyd's
Model: L811
Format: VHS
Type: Tabletop
General Specifications:

Number of Video Heads: 2
Additional Rotating Heads: 0
Remote: IR
Linear Stereo: No
Hi-Fi Stereo: No
Audio Dub: No
Timer (Day/Event): 21/4
HQ Circuits: No

Maintenance Points:

Loading worm gear assembly can become misadjusted, prompting loading and eject malfunctions. If cassette is in unit, tape jam can occur. Re-adjust assembly and lubricate. Also, poor solder joints on end-of-tape sensors can cause auto-stop during recording and playback.

VCR REFERENCE GUIDE

Brand: Lloyd's
Model: L838
Format: VHS
Type: Tabletop
General Specifications:
    Number of Video Heads: 2
    Additional Rotating Heads: 0
    Remote: IR
    Linear Stereo: No
    Hi-Fi Stereo: No
    Audio Dub: No
    Timer (Day/Event): 21/4
    HQ Circuits: No
Maintenance Points:
    Spindles might become loose, causing streaks in picture or total loss of tracking. Carefully readjust the spindles until the picture is clear.

Brand: Magnavox
Model: VR 8510
Format: VHS
Type: Tabletop
General Specifications:
    Number of Video Heads: 2
    Additional Rotating Heads: None
    Remote: Wired
    Linear Stereo: No
    Hi-Fi Stereo: No
    Audio Dub: No
    Timer (Day/Event): 14/2
    HQ Circuits: No
Maintenance Points:
    Worn rubber belt and drive assembly can cause playback and record malfunctions or internal tape jams and spillage.
Manufacturer Cross-Reference:
    Model VR8520 same but with wireless remote.

Brand: Mitsubishi
Model: HS-319UR
Format: VHS
Type: Tabletop

(continued)

General Specifications:
>Number of Video Heads: 3
>Additional Rotating Heads: None
>Remote: IR
>Linear Stereo: No
>Hi-Fi Stereo: No
>Audio Dub: No
>Timer (Day/Event): 14/8
>HQ Circuits: No

Maintenance Points:
>Spindles can become loose, causing lines in top or bottom of screen or total mistracking. Carefully adjust the spindles until the picture is clear.

Manufacturer Cross-Reference:
>Similar to model HS-329UR.

Brand: Multi-Tech
Model: MP-080
Format: 2
Type: VHS
General Specifications:
>Number of Video Heads: 2
>Additional Rotating Heads: None
>Remote: IR
>Linear Stereo: No
>Hi-Fi Stereo: No
>Audio Dub: No
>Timer (Day/Event): 14/4
>HQ Circuits: No

Maintenance Points:
>Take-up reel sensor resistor (usually 1.5 K) out of tolerance, causing auto-stop during playback and record. Replace with lower resistance resistor.

Brand: NEC
Model: VC-N70EV
Format: Beta
Type: Tabletop
General Specifications:
>Number of Video Heads: 4
>Additional Rotating Heads: None
>Remote: IR
>*(continued)*

Linear Stereo: No
Hi-Fi Stereo: Yes
Audio Dub: Yes
Timer (Day/Event): 21/8
HQ Circuits: N/A
Maintenance Points:
Tension lever can become loose, causing unstable tape speed during recording and playback.
Manufacturer Cross-Reference:
Models VC-N56 and VC-N65 lack MTS; VC-N65 lacks audio dub.

Brand: NEC
Model: N-895EU
Format: VHS
Type: Tabletop
General Specifications:
Number of Video Heads: 4
Additional Rotating Heads: 2 (audio)
Remote: IR
Linear Stereo: Yes
Hi-Fi Stereo: Yes
Audio Dub: Yes
Timer (Day/Event): 14/8
HQ Circuits: No
Maintenance Points:
Worn or broken loading belt can prevent tape from loading or unloading. Replace the 1¾-inch loading belt if required.

Brand: Panasonic
Model: PV-1222
Format: VHS
Type: Tabletop
General Specifications:
Number of Video Heads: 2
Additional Rotating Heads: None
Remote: N/A
Linear Stereo: No
Hi-Fi Stereo: No
Audio Dub: No
Timer (Day/Event): 7/1
HQ Circuits: No
(continued)

Maintenance Points:

Rewind and fast-forward gear assembly can become jammed (lubricate to clear) causing sluggish rewind and fast-forward operations.

Manufacturer Cross-Reference:

Panasonic VCRs are made by Matsushita Electric.

Brand: Panasonic
Model: PV-1340
Format: VHS
Type: Tabletop
General Specifications:

Number of Video Heads: 2
Additional Rotating Heads: None
Remote: IR
Linear Stereo: No
Hi-Fi Stereo: No
Audio Dub: No
Timer (Day/Event): 14/4
HQ Circuits: No

Maintenance Points:

Worn or dirty idler tire can cause internal tape spillage.

Manufacturer Cross-Reference:

Panasonic products are produced exclusively by Matsushita.

Brand: Panasonic
Model: PV-1462
Format: VHS
Type: Tabletop
General Specifications:

Number of Video Heads: 2
Additional Rotating Heads: 2 (audio)
Remote: IR
Linear Stereo: Yes
Hi-Fi Stereo: Yes
Audio Dub: No
Timer (Day/Event): 14/4
HQ Circuits: Yes

Maintenance Points:

End-of-tape sensor LED can burn out, triggering auto shutoff during recording and playback. Also, a worn or dirty idler tire can cause internal tape spillage and auto-shutoff.

(continued)

VCR REFERENCE GUIDE

Manufacturer Cross-Reference:

Model PV-1442 has MTS; model PV-1461 has MTS and linear stereo (no hi-fi). Panasonic VCRs are made by Matsushita Electric.

Brand: Panasonic
Model: PV-1540
Format: VHS
Type: Tabletop
General Specifications:
　　Number of Video Heads: 4
　　Additional Rotating Heads: None
　　Remote: IR
　　Linear Stereo: No
　　Hi-Fi Stereo: No
　　Audio Dub: No
　　Timer (Day/Event): 14/4
　　HQ Circuits: Yes

Maintenance Points:

Mechanical switches can become dirty, causing intermittent audio and video problems. In many cases, dirty audio switches cause recording of picture but not sound. Loading mechanism can become misaligned, causing loading and unloading problems. Re-alignment can be done manually or with an alignment jig. Also, a worn or dirty idler tire can cause playback problems (tape might not play at all or could become jammed in mechanism).

Manufacturer Cross-Reference:

Model PV-1542 stereo model but no HQ, models PV-1562, 1563, and 1564 stereo, stereo/mts, and stereo hi-fi/MTS (respectively). Panasonic VCRs are made by Matsushita Electric.

Brand: PortaVideo
Model: VP-3100
Format: VHS
Type: Portable
General Specifications:
　　Number of Video Heads: 2
　　Additional Rotating Heads: None
　　Remote: N/A
　　Linear Stereo: No
　　Hi-Fi Stereo: No
　　Audio Dub: No
　　Timer (Day/Event): N/A
　　HQ Circuits: No

(continued)

Maintenance Points:

Loading mechanism can become misaligned, causing tape eject to malfunction. Re-align manually or with alignment jig.

Brand: Quasar
Model: VH-5162
Format: VHS
Type: Tabletop
General Specifications:

Number of Video Heads: 2
Additional Rotating Heads: None
Remote: IR
Linear Stereo: No
Hi-Fi Stereo: No
Audio Dub: No
Timer (Day/Event): 14/4
HQ Circuits: Yes

Maintenance Points:

Loading and cassette lift mechanism can become misaligned, requiring realignment. Idler tire can become dirty, glazed, or worn, causing playback and rewind difficulties. Clean or replace as required.

Manufacturer Cross-Reference:

Quasar products are manufactured by Matsushita; most are similar to Panasonic models. Model VH5163 similar.

Brand: Quasar
Model: VH 5260
Format: VHS
Type: Tabletop
General Specifications:

Number of Video Heads: 4
Additional Rotating Heads: None
Remote: IR
Linear Stereo: No
Hi-Fi Stereo: No
Audio Dub: No
Timer (Day/Event): 14/4
HQ Circuits: Yes

Maintenance Points:

Worn idler tire causes rewind and play malfunctions. Excessively worn rubber may cause internal tape spillage.

(continued)

VCR REFERENCE GUIDE

Manufacturer Cross-Reference:
Quasar products are manufactured by Matsushita; most are similar to Panasonic models. Model VH5268 similar, but has linear stereo; model VH5262 similar, also mono.

Brand: Quasar
Model: VH 5260/61
Format: VHS
Type: Tabletop
General Specifications:
Number of Video Heads: 4
Additional Rotating Heads: None
Remote: IR
Linear Stereo: No
Hi-Fi Stereo: No
Audio Dub: No
Timer (Day/Event): 14/4
HQ Circuits: Yes
Maintenance Points:
Bad servo controller integrated circuit can cause the machine to run by itself and ignore front panel controls.
Manufacturer Cross-Reference:
Quasar products are manufactured by Matsushita; most are similar to Panasonic models.

Brand: Quasar
Model: VH5665
Format: VHS
Type: Tabletop
General Specifications:
Number of Video Heads: 4
Additional Rotating Heads: 2 (audio)
Remote: IR
Linear Stereo: Yes
Hi-Fi Stereo: Yes
Audio Dub: No
Timer (Day/Event): 21/8
HQ Circuits: Yes
Maintenance Points:
Worn idler tire can cause malfunction in play and rewind functions (tire might hold for fast-forward).

(continued)

278

Manufacturer Cross-Reference:

Quasar products are manufactured by Matsushita; most are similar to Panasonic models. Model VH5865 similar but has additional features on remote and more cable channels.

Brand: Radio Shack/Realistic
Model: Model 18
Format: VHS
Type: Tabletop
General Specifications:
  Number of Video Heads: 2
  Additional Rotating Heads: None
  Remote: IR
  Linear Stereo: No
  Hi-Fi Stereo: No
  Audio Dub: No
  Timer (Day/Event): 14/4
  HQ Circuits: No
Maintenance Points:

Burned out 1 K resistor (R634) for take-up sensor can cause auto-shutoff after short period of operation.

Manufacturer Cross-Reference:

Radio Shack VCRs are manufactured by Sanyo Tokyo; Model 17 includes HQ.

Brand: RCA
Model: TGP-1500
Format: VHS
Type: Tabletop
General Specifications:
  Number of Video Heads: 2
  Additional Rotating Heads: None
  Remote: N/A
  Linear Stereo: No
  Hi-Fi Stereo: No
  Audio Dub: No
  Timer (Day/Event): 1/1
  HQ Circuits: No
Maintenance Points:

Mechanical tuner can become dirty, necessitating cleaning. Heavy use of rubber tires and belts; recondition or replace every 3 to 4 years to avoid tape loading, threading, and playback problems. Since about 1983-1984, RCA models are built by Hitachi.

VCR REFERENCE GUIDE

Brand: RCA
Model: VDT-501
Format: VHS
Type: Tabletop
General Specifications:
    Number of Video Heads: 2
    Additional Rotating Heads: None
    Remote: N/A
    Linear Stereo: No
    Hi-Fi Stereo: No
    Audio Dub: No
    Timer (Day/Event): 1/1
    HQ Circuits: No
Maintenance Points:
    Burned out end-of-tape sensor can cause auto-stop during recording and playback. Since about 1983-1984, RCA models are built by Hitachi.

Brand: RCA
Model: VFP-170
Format: VHS
Type: Tabletop
General Specifications:
    Number of Video Heads: 2
    Additional Rotating Heads: None
    Remote: Wired
    Linear Stereo: No
    Hi-Fi Stereo: No
    Audio Dub: No
    Timer (Day/Event): 1/1
    HQ Circuits: No
Maintenance Points:
    Belts and tires can become dirty or worn, and cause auto-shut off during recording and playback. Impedance rollers can become cracked, causing erratic playback speeds and poor recordings.
Manufacturer Cross-Reference:
    Since about 1983-1984, RCA models are built by Hitachi.

Brand: RCA
Model: VJP-900
Format: VHS
Type: Portable
General Specifications:
Number of Video Heads: 5
Additional Rotating Heads: None
Remote: IR
Linear Stereo: Yes
Hi-Fi Stereo: No
Audio Dub: Yes
Timer (Day/Event): ??
HQ Circuits: No
Special Features: Docking design allows portable to "plug" into base unit.
Maintenance Points:
Worn or dirty idler tire causes record and playback malfunctions. Can also prompt tape spillage.
Manufacturer Cross-Reference:
Made by Hitachi; same as model VT-7P.

Brand: RCA
Model: VLT-375
Format: VHS
Type: Tabletop
General Specifications:
Number of Video Heads: 2
Additional Rotating Heads: None
Remote: Wired
Linear Stereo: No
Hi-Fi Stereo: No
Audio Dub: No
Timer (Day/Event): 14/4
HQ Circuits: No
Maintenance Points:
Worn or dirty idler tire causes record and playback malfunctions. Can also prompt tape spillage.
Manufacturer Cross-Reference:
Made by Hitachi.

VCR REFERENCE GUIDE

Brand: Samsung
Model: VT-210
Format: VHS
Type: Tabletop
General Specifications:
> Number of Video Heads: 2
> Additional Rotating Heads: None
> Remote: Wired
> Linear Stereo: No
> Hi-Fi Stereo: No
> Audio Dub: No
> Timer (Day/Event): 7/1
> HQ Circuits: No

Maintenance Points:
> Worn, glazed, or dirty idler tire may cause malfunctions in rewind, fast-forward, or play modes and can affect accuracy of the tape counter.

Brand: Sanyo
Model: VCR4300
Format: Beta
Type: Tabletop
General Specifications:
> Number of Video Heads: 2
> Additional Rotating Heads: None
> Remote: Wired
> Linear Stereo: No
> Hi-Fi Stereo: No
> Audio Dub: No
> Timer (Day/Event): 3/1
> HQ Circuits: N/A

Maintenance Points:
> Mechanical tuner can become dirty and require cleaning and adjustment (only certain channels will be affected, but clean contacts for all channels).

Brand: Sanyo
Model: VCR4400
Format: Beta
Type: Tabletop
General Specifications:
> Number of Video Heads: 2
> Additional Rotating Heads: None
> Remote: Wired
> Linear Stereo: No
> *(continued)*

Hi-Fi Stereo: No
Audio Dub: No
Timer (Day/Event): 3/1
HQ Circuits: N/A
Maintenance Points:
Worn, glazed, or dirty idler wheel can cause picture irregularities and tape jamming. Tape might skip and picture might go blank for a moment; clean all parts in the tape path and adjust the spindles if they have worked loose.

Brand: Sanyo
Model: VCR4750
Format: Beta
Type: Tabletop
General Specifications:
Number of Video Heads: 3
Additional Rotating Heads: None
Remote: IR
Linear Stereo: No
Hi-Fi Stereo: No
Audio Dub: No
Timer (Day/Event): 14/8
HQ Circuits: N/A
Maintenance Points:
Worn, glazed, or dirty idler tire can induce auto-stop; pushing PLAY button has no effect and machine might eject the tape.

Brand: Sanyo
Model: VCR7200
Format: Beta
Type: Tabletop
General Specifications:
Number of Video Heads: 2
Additional Rotating Heads: None
Remote: IR
Linear Stereo: No
Hi-Fi Stereo: Yes
Audio Dub: No
Timer (Day/Event): 14/8
HQ Circuits: N/A
Maintenance Points:
Worn, glazed, or dirty idler tire can induce auto-stop; pushing PLAY button has no effect and machine might eject the tape.

VCR REFERENCE GUIDE

Brand: Sears
Model: 53091
Format: Beta
Type: Tabletop
General Specifications:
    Number of Video Heads: 2
    Additional Rotating Heads: None
    Remote: None
    Linear Stereo: No
    Hi-Fi Stereo: No
    Audio Dub: No
    Timer (Day/Event): 3/1
    HQ Circuits: N/A
Maintenance Points:
    Worn idler tire slips, causing mis-threading, tape jamming, and cassette loading problems.

Brand: Sharp
Model: VC-489U
Format: VHS
Type: Tabletop
General Specifications:
    Number of Video Heads: 4
    Additional Rotating Heads: 2 (audio)
    Remote: IR
    Linear Stereo: Yes
    Hi-Fi Stereo: Yes
    Audio Dub: Yes
    Timer (Day/Event): 14/5
    HQ Circuits: No
Maintenance Points:
    Slipping belts and rollers can cause video problems (usually colored lines on top or bottom of screen) and threading malfunctions.

Brand: Sony
Model: SL2300
Format: Beta
Type: Tabletop
General Specifications:
    Number of Video Heads: 2
    Additional Rotating Heads: None
    Remote: Wired
    (continued)

Linear Stereo: No
Hi-Fi Stereo: No
Audio Dub: No
Timer (Day/Event): 7/1
HQ Circuits: N/A

Maintenance Points:
Stripped or worn loading gears can cause tape loading, unloading, and threading difficulties. If cassette is inside the VCR when the problem occurs, the tape can jam.

Brand: Sony
Model: SL2410
Format: Beta
Type: Tabletop
General Specifications:
Number of Video Heads: 2
Additional Rotating Heads: None
Remote: IR
Linear Stereo: No
Hi-Fi Stereo: No
Audio Dub: No
Timer (Day/Event): 21/4
HQ Circuits: N/A

Maintenance Points:
Stripped loading gears can cause tape loading, unloading, and threading difficulties. If cassette is inside the VCR when the problem occurs, the tape can jam.

Brand: Sony
Model: SL2500
Format: Beta
Type: Tabletop
General Specifications:
Number of Video Heads: 2
Additional Rotating Heads: 0
Remote: Wired
Linear Stereo: No
Hi-Fi Stereo: No
Audio Dub: No
Timer (Day/Event): 7/1
HQ Circuits: N/A

Maintenance Points:
Gears in loading mechanism can become jammed and break, causing loading and unloading malfunctions. Maladjusted loading cassette-in and tape-thread switches can cause machine to auto-stop after 3 to 5 seconds.

VCR REFERENCE GUIDE

Brand: Toshiba
Model: V-M50
Format: Beta
Type: Tabletop
General Specifications:
    Number of Video Heads: 2
    Additional Rotating Heads: None
    Remote: Wired
    Linear Stereo: No
    Hi-Fi Stereo: No
    Audio Dub: No
    Timer (Day/Event): 7/1
    HQ Circuits: N/A
Maintenance Points:
    Loading and lift mechanism can become worn and broken, requiring replacement. Loading gears might jam due to dirty and grime build-up, causing groaning during cassette loading and unloading (in some cases, cassette jams). Clean and lubricate the gear system.

Brand: Zenith
Model: VR 1820
Format: VHS
Type: Tabletop
General Specifications:
    Number of Video Heads: 4
    Additional Rotating Heads: None
    Remote: IR
    Linear Stereo: No
    Hi-Fi Stereo: No
    Audio Dub: No
    Timer (Day/Event): 14/4
    HQ Circuits: Yes
Maintenance Points:
    Wavy, fuzzy picture can result from tape not fully threaded onto drum because of dirty threading tracks and worn idler tire. Clean and lubricate threading tracks and recondition or replace idler tire.
Manufacturer Cross-Reference:
    Zenith products are predominantly made by JVC. Model VR2100 is similar but lacks HQ and has some additional features.

11

Camcorders

Camcorders are combination video cameras and VCRs slimmed down to a totable size and weight. Camcorders contain a fully functional VCR that requires much of the same maintenance and repair as a full-sized deck. You must exercise caution, however, when working on camcorders, because their small profile makes even routine maintenance and repair difficult.

DISASSEMBLY

When at all possible, avoid disassembling your camcorder. Most camcorder models use "hidden"screws as a measure to discourage users from taking them apart (most still-camera manufacturers have used this scheme for years). The design of camcorders necessarily entails miniaturization and special construction techniques. Many parts are held in place by pressure or springs, and opening the case of the camcorder might release these tiny parts. However, if you must disassemble the camcorder, follow these steps:

➡ If at all possible, obtain a copy of the service manual. The manual will provide instructions on how to disassemble the unit and contains tips for keeping all the parts together. Finally, the manual contains an assembly or exploded diagram that helps when putting the camcorder back together again.
➡ Lacking a service manual, disassemble the camcorder by carefully removing the exposed screws on the exterior of the case. If loosening a screw doesn't seem to loosen the case, temporarily retighten it and try another. Some screws

on the exterior of the camcorder serve to anchor internal parts, and these should not be removed.

➡ Screws are often hidden under the peel-away operating panels or the vinyl cushioning used on most camcorder models. Carefully peel away the panel or cushion to expose these hidden screws. Look also in or on the grip or battery compartment.

➡ Place the screws and all other parts you remove from the camcorder in a tray. Keep the parts separate to facilitate reassembly. It is critical that you reassemble the camcorder using the same parts. Don't mix up the screws, washers, and other hardware. If necessary, take notes during disassembly.

➡ When all the screws are removed, carefully lay the camcorder on the table and separate the outer casing (or cover) from the main body. Go slowly; note the location of all parts that come loose.

➡ Some camcorders require you to loosen or remove the tapeloading cover before the case can be completely removed. The cover is generally held in place by two screws.

➡ Wires might attach the camcorder body to the case (particularly at the battery and operating control panels). Do not strain these wires by pulling the case too far from the camcorder. If wires tether the case to the camcorder, see if they can be removed at the connections. If not, carefully lay the case beside the camcorder body and work around it.

HEAD CLEANING

Because of the difficulty in disassembling and reassembling a camcorder, it's often better to clean the heads using a cassette cleaner. Just about any wet or wet/dry cleaner will do; experiment with a few brands until you find a cleaning system that works best.

Use the head cleaning tape as recommended by the manufacturer. After using the tape, wait several minutes for the cleaning fluid that has been deposited inside the camcorder to evaporate completely. Inserting and playing a tape when the unit is still damp can re-clog the heads and can cause the tape to stick.

A more thorough head cleaning entails removing the cover of the camcorder and using swabs dipped in alcohol or other cleaning agent. Follow the head cleaning procedures found in Chapter 5. Note that compact VHS-C camcorders have four or more video heads; be sure to clean them all. Many 8mm and some VHS camcorders also have flying erase heads built onto the head drum. Be sure to clean these as well.

All camcorders are the top-load variety, which means that when the tape-loading cover is open (the unit is ready to accept a tape), the video head drum is somewhat accessible. If possible, keep the case of the camcorder on and clean the video heads by using a sponge- or chamois-tipped swab, poking it through the opening in the tape-loading cover. Be sure that you don't apply excessive pressure on the heads. You can rotate the drum to access all of the heads by spinning the top part with your fingers or the eraser end of a pencil. Take care not to touch the heads with any instrument but the swab soaked in cleaning fluid.

DUSTING AND GENERAL CLEANING

Because camcorders are primarily used out of doors, they are especially likely to become soiled by dust, dirt, and other contaminants. At regular intervals, clean the exterior of the camcorder with a mild household cleaner, such as 409 or Fantastik, but remember to apply the cleaner to the sponge or towel—never spray the cleaner directly onto the camcorder as the excess might ooze inside.

Clean around the control switches with a cotton swab dipped in cleaner or alcohol. Avoid the use of petroleum-based solvent cleaners, especially acetone. These can remove paint and melt the plastic casing of the camcorder.

If you have disassembled the camcorder, use a soft brush to lightly dust the interior of the unit. A can of compressed air can be used to rid dust and dirt from hard-to-reach places. If the sediment is heavy, use a cleaner/degreaser, as described in Chapter 5.

Parts Cleaning and Replacement

Camcorders use slightly fewer parts than their full-sized cousins, but that doesn't mean that mechanical failure is less likely to occur. Rubber belts, rollers, and other components can fail after time and might need cleaning or replacement.

Generally, you clean and replace parts such as idlers, belts, and pressure rollers the same as you do regular VCRs. Because of their compact nature, however, a variety of novel schemes are used to attach components to the camcorder.

Tires and rollers might be secured with just a press fit, but the fit might be extremely tight, requiring a special tool for removal and replacement. On other camcorders, rollers and tires are secured using miniature jeweler's screws or hex screws. Be sure to use the proper size tool or you risk stripping the head of the screws completely. If some screws seem overly tight, apply a *very* small dab of oil to the head and shaft area and wait for the oil to penetrate.

It's a joy to work on a camcorder that is designed so that drive belts are out in the open and are readily accessible. However, the midget design of camcorders often precludes this, and belts are typically hidden under other components and are difficult to reach.

At least some portion of the belt is usually accessible, and you can use a pair of tweezers, surgical forecepts, or dentists picks to examine and—only if necessary—extract the belt for replacement. Note that removing the belt is the easy part; putting a new one in is only slightly less frustrating as getting the ball in the milk bottles at the county fair. Note that some components, such as motors and circuit boards, need to be removed in order to replace a belt.

CAMCORDERS

CAMERA CARE

The camera portion of the camcorder needs little maintenance beyond the following:

→ Clean the lens prior to each use. Never clean the lens with a dry lens tissue; wet the tissue or lens with a suitable cleaner. If you have attached to a clear UV or Skylight filter over the front of the lens (to protect the actual camera lens from accidental injury), periodically remove the filter and clean it.

→ Use a cotton swab dipped in alcohol to clean the lens barrel and camera controls. Blot dry.

→ If the camera gets out of whack—the focus is thrown off or the image appears lop-sided, for example—adjustment of some internal controls might be necessary. Do not attempt adjustment of any internal controls without the assistance of the service manual. The manual provides test points and procedures for aligning the camera section for proper picture-taking. Note that the latest camcorders use solid-state imaging devices, which rarely require the adjustments needed by vacuum-tube pickups.

Appendix A

Sources

Akai Division
Mitsubishi Electric Sales America
225 Old New Brunswick Rd.
Piscataway, NJ 08854

Aiwa America
35 Oxford Dr.
Moonachie, NJ 07074

Allsop
4201 Meridian Box 23
Bellingham, WA 98227

Canon U.S.A.
1 Canon Plaza
Lake Success NY 11042

Chinon America, Inc.
43 Fadem Rd.
Springfield, NJ 07081

Curtis Mathes Corp.
144 Greenway Dr.
Irving, TX 75062

Discwasher
1407 No. Providence Rd.
PO Box 6021
Columbia, MO 65205

Eastman Kodak Co.
343 State St.
Rochester, NY 14650

Elmo Manufacturing Corp.
70 New Hyde Park Rd.
New Hyde Park, NY 11040

Emerson Radio Corp.
One Emerson Lane
North Bergen, NJ 07047

Fisher Corp.
21314 Lassen St.
Chatsworth, CA 91311

Fuji
350 5th Ave.
New York, NY 10118

APPENDIX A: SOURCES

General Electric
One Wellner Dr.
Portsmouth, VA 23705

Goldstar Electronics Inc.
1050 Wall St.
Lyndhurst, NJ 07071

Hitachi Denshi
175 Crossways Park West
Woodbury, NY 11797

Instant Replay
2951 South Bayshore Dr.
Miami, FL 33133

JC Penney
1339 Tolland Turnpike
Manchester, CT 06040-1567

JVC Company of America
41 Slater Dr.
Elmwood Park, NJ 07407

Kenwood Electronics
1315 Watson Center Rd.
Carson, CA 90745

Kyocera
411 Settle Dr.
Paramus, NJ 07652

Matsushita Electric Corp. of America
1 Panasonic Way
Secaucus, NY 07094

Minolta Corp.
101 Williams Dr.
Ramsey, NJ 07446

Mitsubishi Electric Sales of America
5757 Plaza Dr.
PO Box 6007
Cypress, CA 90630-0007

Montgomery Ward
535 W. Chicago Ave.
Chicago, IL 60610

NAP Consumer Electronics (Magnavox)
Interstate 40 & Straw Plains Pike
PO Box 6950
Knoxville, TN 37914-1810

NEC Home Electronics USA
1255 Michael Dr.
Wood Dale, IL 60191-1094

Nikon
623 Stewart Ave.
Garden City, NY 11530

Olympus Corp.
Crossways Park
Woodbury, NY 11797

Panasonic Co.
One Panasonic Way
Secaucus, NJ 07094

Pentax Corp.
35 Iverness Dr. East
Englewood, CO 80112

Philco
Box 14810
Knoxville, TN 37914

Pioneer Video, Inc.
200 West Grand Ave.
Montvale, NJ 07645

Portavideo
1930 W. Third
Box 22130
Tempe, AZ 85282

Quasar Company
1325 Pratt Blvd.
Elk Grove Village, IL 60007

RCA Consumer Electronics Business
600 North Sherman Dr.
Indianapolis, IN 46201

Radio Shack/Realistic
1800 One Tandy Center
Fort Worth, TX 76102

Ricoh
5 Dedrick Place
West Caldwell, NJ 07006

Samsung Electronics America, Inc.
301 Mayhill St.
Saddle Brook, NJ 07662

Sansui Electronics Corp.
1250 Valley Brook Ave.
Lyndhurst, NJ 07071

Sanyo Electric, Inc.
1200 W. Artesia Blvd.
Compton, CA 90220

Sears, Roebuck & Co.
4640 Roosevelt Blvd.
Philadelphia, PA 19132

Sharp Electronics Corp.
Sharp Plaza
Mahwah, NJ 07430

Sony Corp of America
Sony Drive
Park Ridge, NJ 07656

Supra USA
10-27 45th Ave.
Long Island City, NY 11101

Sylvania/NAP Consumer Electronics
Interstate 40 & Straw Plains Pike
PO Box 6950
Knoxville, TN 37914

Technics/Panasonic
One Panasonic Way
Secaucus, NJ 07094

Teknika Electronics Corp.
353 Rte. 46 West
Fairfield, NJ 70006

Toshiba America Inc.
82 Totowa Rd.
Wayne, NJ 07470

Vector Research
20600 Nordoff St.
Chatsworth, CA 91311

Vivitar
1630 Stewart St.
Santa Monica, CA 90406

Zenith Electronics Corp.
1000 Milwaukee Ave.
Glenview, IL 60025

Appendix B

Further Reading

Interested in learning more about video and VCRs? Here is a selected list of magazines and books that can enrich your understanding and enjoyment of the world of video.

MAGAZINES

Audio/Video Buyer's Guide
HARRIS PUBLICATIONS
1115 BROADWAY
NEW YORK, NY 10010
   Seasonal buyers' guide on home and car audio and video, available at newsstands. Light reading but somewhat out-of-date by the time you get it. No subjective reviews of VCR hardware.

Modern Electronics
76 NORTH BROADWAY
HICKSVILLE, NY 11801
   Monthly magazine for electronics hobbyists. Occasional technical article on video.

Radio-Electronics
500 BI-COUNTY BLVD.
FARMINGDALE, NY 11735

Monthly magazine for electronics hobbyists. Occasional technical article on VCRs and video, plus some buyer's guides on VCRs and other hardware.

VIDEO
460 W. 34TH STREET
NEW YORK, NY 10001
Monthly magazine on all things video. Includes hardware and software reviews, technical articles (how it works), buyer's guides, and how-to stories. *VIDEO* also regularly publishes special buyer's guide issues.

Video Review
902 BROADWAY
NEW YORK, NY 10010
Monthly magazine for video enthusiasts. Although the slant is decidedly video software, the magazine does carry regular articles on the latest trends in video. Also includes hardware buyer's guides and hands-on test reports.

Videofax
PO BOX 481248
LOS ANGELES, CA 90048-9743.
Quarterly newsletter/magazine devoted to hard-core video enthusiasts. Highly recommended. The articles are informative, highly biased, and sometimes overly critical, but if you are looking to buy video gear or get more out of the stuff you have, you ought to check out current and back issues of Videofax. Sold mainly by subscription, although some larger newsstands carry it.

BOOKS

How to Read Electronic Circuit Diagrams—2nd Ed., ROBERT M. BROWN, PAUL LAWRENCE, AND JAMES A. WHITSON
TAB BOOKS, CATALOG #2880
How to read and interpret schematic diagrams.

Camcorder Handbook, The, GERALD V. QUINN
TAB BOOKS, CATALOG #2801
Backgrounder and how-to on camcorders. Includes formats, how they work, and buyer's guidance.

Digital Electronics Troubleshooting—2nd Ed., JOSEPH J. CARR
TAB BOOKS, CATALOG #2750
Theory and practice of troubleshooting digital circuits.

APPENDIX B: FURTHER READING

Handbook of Electronic Safety Procedures, EDWARD A. LACY
TAB BOOKS, CATALOG #1420

Covers safety precautions and procedures when troubleshooting and repairing electric and electronic devices.

How to Troubleshoot & Repair Electronic Circuits, ROBERT L. GOODMAN
TAB BOOKS, CATALOG #1218
General troubleshooters guide; both analog and digital.

Meters and Scopes, How to Use Test Equipment, ROBERT J. TRAISTER
TAB BOOKS, CATALOG #2826
How to properly use oscilloscopes, volt ohmmeters, and other electronic testing gear.

Understanding Digital Electronics, R. H. WARRING
TAB BOOKS, CATALOG #1593
Introduction to the principles of digital theory.

Appendix C
Maintenance Log

Use the maintenance log on the following pages to keep track of the routine upkeep and servicing you do on your video cassette recorder. Feel free to make photocopies of these pages and keep the copies with your deck's instruction manual. Maintenance logs are of little use unless you stick to them religiously, so make sure you write down anything and everything you do to your deck.

Maintenance Log

Video Cassette Recorder Maintenance and Repair Log

VCR Vital Statistics

Brand_____

Model_____

Date Purchased_____Serial Number_____

Where Purchased_____

Sales Person_____Warranty Period_____

New_____Used_____ If Used, How Old When Purchased?_____

Type: AC Operated Home_____Portable_____Camcorder_____

VCR Design

Number of Video Heads: 2_____ 3_____ 4_____ 5_____

Format: VHS_____Beta_____8mm_____Other_____

Audio: Stereo_____Hi-Fi_____MTS Ready_____MTS Adapt_____Dub_____

Special Circuits: Super Beta_____Super VHS_____HQ_____Digital Effects_____

PM Schedule Supplies (Type or Brand and Source)

Household Spray Cleaner_____

Non Petroleum-Based Solvent Cleaner_____

Oil Lubricant_____

Grease Lubricant_____

Electrical Contact Cleaner_____

Compressed Air_____

Rubber Cleaner_____

Other:_____

_____

_____

_____

_____

_____

_____

_____

Tools (List)

_____

_____

_____

Page 2   Video Cassette Recorder Maintenance and Repair Log   Page 2

General Maintenance

Date				
Clean PCB				
Clean Heads				
Clean Transport				
Clean Controls				
Check Operation				
Inspect Wiring				

Lubrication

Date				
Loading Assembly				
Threading Assembly				
Transport Assembly				
Other_____				

Contact Cleaning

Date				
Power Swich				
Front Panel Switches				
Interlock Switches				
Remote Switches				
Remote Battery Terminals				
Connectors				
Head Ass'y				

APPENDIX C: MAINTENANCE LOG

Video Cassette Recorder Maintenance and Repair Log

Repair/Replacement

Date				
Capstan Belt				
Idler Tire				
Idler Drive Belt				
End-of-Tape Sensor/Lamp				
Switch				
Capstan Pinch Roller				
Capstan Motor				
Main PCB				
Video Head Assembly				

Problem Occurrence

Date_____
Description of Problem_____
_____
Action Taken_____
_____
Meter Readings, Measurements, Etc._____
_____

Date_____
Description of Problem_____
_____
Action Taken_____
_____
Meter Readings, Measurements, Etc._____
_____

Page 4   Video Cassette Recorder Maintenance and Repair Log   Page 4

Problem Occurrence (Continued)

Date_____
Description of Problem_____
_____
_____
Action Taken_____
_____
_____
Meter Readings, Measurements, Etc._____
_____

Date_____
Description of Problem_____
_____
_____
Action Taken_____
_____
_____
Meter Readings, Measurements, Etc._____
_____

Date_____
Description of Problem_____
_____
_____
Action Taken_____
_____
_____
Meter Readings, Measurements, Etc._____
_____

Special Notes
_____
_____
_____
_____
_____

Appendix D

Soldering Tips
and Techniques

Successful repair of your VCR depends largely on how well you can solder two wires together. Soldering sounds and looks simple enough, but there really is a science about it. If you are unfamiliar with soldering or want a quick refresher course, read this short soldering primer.

TOOLS AND EQUIPMENT

Good soldering requires the proper tools. If you don't have them already, they can be purchased at Radio Shack or most any electronics store.

Soldering Iron

You'll need a soldering iron, of course, but not just any old soldering iron. Get a soldering "pencil" with a low-wattage heating element. For electronics work, the heating element should not be higher than about 30 watts. Most soldering pencils are designed so that you can change heating elements as easy as changing a lightbulb.

DO NOT use the instant-on type soldering guns, favored in the old tube days. They create far too much unregulated heat, and they are too large to effectively solder most joints on a PCB.

If your soldering iron has a temperate control and readout, dial it to between 665 and 680 degrees. This provides maximum heat with the minimum danger of damage to the electronic components. When you are not using your soldering iron, keep it in an insulated stand. Don't rest the iron in an ashtray or precariously on the carpet. You or some precious belonging is sure to be burned.

302

APPENDIX D: SOLDERING TIPS AND TECHNIQUES

Soldering Tip

The choice of soldering tip is important. For best results, use a fine tip designed specifically for printed circuit board use. Tips are made to fit certain types and brands of heating elements, so make sure you get the kind for your iron. If the tip doesn't come pre-tinned, tin it by attaching the tip to the iron and heating it up. After the iron is hot, apply a thin coat of solder to the entire tip.

Sponge

Keep a damp sponge by the soldering station and use it to wipe off extra solder. You'll have to re-wet the sponge now and then if you are doing lots of soldering.

Solder

You should use only rosin core solder. It comes in different thicknesses; for best results, use the thin type (0.050″) for PCB work. NEVER use acid core or silver solder on electronic equipment.

Soldering Tools

Basic soldering tools include a good pair of small needle-nose pliers, tweezers, wire strippers and wire cutters (sometimes called side or diagonal cutters). The stripper should have a dial that lets you select the gauge of wire you are using. A pair of "nippy" cutters that cut wire leads flush to the surface of the board are handy but not absolutely essential.

Cleaning Supplies

After soldering and when the components and board are cool, you should spray or brush on some flux remover. Isopropyl alcohol can also be used for cleaning.

Solder Vacuum

A solder vacuum is a suction device that is used to pick up excess solder. It is often used when desoldering—removing a wire or component from the board. Solder can also be removed using a length of copper braid. Most electronics stores sell spools of it specifically for solder removal.

BASIC SOLDERING

The basis of successful soldering is that the soldering iron is used to heat up the work, whether it be a component lead, a wire, or whatever. You then apply the solder to the work. DO NOT apply solder directly to the soldering iron. If you take the shortcut by melting the solder on the iron, you might end up with a "cold" solder joint. A cold joint doesn't adhere well to the metal surfaces of the part or board, so electrical connection is impaired.

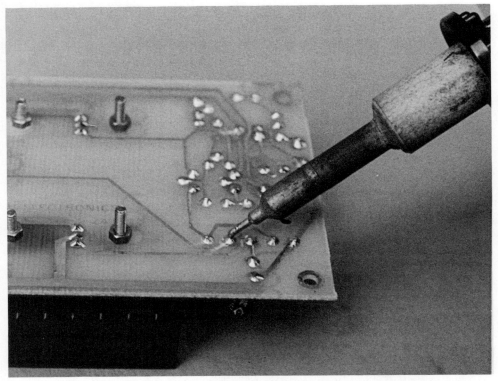

Fig. D-1. The proper technique for applying the soldering iron to a printed-circuit board.

Once the solder flows around the joint (and some will flow to the tip), remove the iron and let the joint cool. Avoid disturbing the solder as it cools; a cold joint might result. Do not apply heat any longer than necessary. Prolonged heat can permanently ruin electronic components. A good rule of thumb is that if the iron is on any one spot for more than five seconds, it's too long.

If at all possible, you should keep the iron at a 30 to 40 degree angle to work, as shown in Fig. D-1. Most tips have a beveled tip for this purpose.

Apply only as much solder to the joint as is required to coat the lead and circuit-board pad. A heavy-handed soldering job might lead to soldering bridges, which is when one joint melds with other joints around it. At best, solder bridges cause the circuit to not work; at worse, they cause short circuits that can burn out the entire board.

REPLACING COMPONENTS

Removing a soldered component first requires that you remove all the solder holding it in place. Use the soldering iron to melt the joint, and as the solder flows, suck it up with a solder vacuum or wick. Remove enough solder so that the component lead is free. If you can't get all the solder up the first time, let the joint cool and try again.

Clean the old joint and the leads of the replacement component in alcohol. This removes any oil that can impair the grip of the joint after soldering.

Insert the new component gently. Don't pull on the lead or you can damage the component. Once the part is seated on the board, bend the leads slightly to keep it in place. Solder as usual. If you are resoldering a wire onto a terminal, wrap the stripped end of the wire around the eyelet of the terminal prior to soldering.

A GOOD SOLDER JOINT

A good solder joint should be bright and shiny. A joint that looks dull is probably cold and should be remade. The joint should not have any sharp "peaks." If so, the solder didn't flow well enough to make a good connection. Remake the joint, being sure to apply the iron to the work and not to the iron. Excess solder that forms on the tip (another cause for the "peaks") should be removed using the damp sponge.

ELECTROSTATIC DISCHARGE

Electrostatic discharge—better known as a "carpet shock"—can ruin electronic components. You should remove the excess static buildup from your body by touching some grounded metal object prior to soldering, or before handling electronic parts and boards. If you are soldering transistors and integrated circuits, you should use a grounded soldering iron as well as an anti-static wrist band and table mat.

IRON TIP MAINTENANCE AND CLEANUP

After soldering, let the iron cool. Loosen the tip from the heating element and store it for next use. After several soldering sessions, the tip should be cleaned using a soft brush. Don't file it or sand it down with emery paper.

Invariably, little nuggets of solder will be left around after a repair job. Make sure that these balls of once-molten solder balls are not left on the PCB or in the VCR cabinet. The solder might bridge wires or board traces together, causing a serious short circuit. Inspect your work carefully and use a soft brush to wisk away stray bits of solder.

Appendix E
Attaching an F-Connector

You can either buy video cables cut to length with the proper F-connector already attached to them, or you can make the cables yourself. Making your own cables saves you money, and the cables can be cut to the proper length for the job. Figure E-1 is a step-by-step guide to attaching an F-connector onto coaxial cable. Please note that F-connectors are designed for a specific cable type, for example, connectors designed for RG59 cable won't fit on RG6 cable.

WHAT TO DO

Step 1: Strip the outer jacket insulation from the cable to expose about 7/16-inch of the braided wire. At this point, don't cut the braid.

Step 2: Fan out the braid, fold it back over the cable jacket, and trim to about 1/8 inch in length.

Step 3: Some cables have an aluminum shield under the braid. Strip off the excess shield and insulation foam to expose 1/4-inch of the center conductor—do not cut too deeply into the dielectric or you'll nick the center conductor. Make sure the center conductor is clean and bright-looking.

Step 4: Slide the F-connector over the foam but *under* the braid. Push the fitting into the cable until it seats firmly and no braid is showing. If the fitting is the crimp type, crimp it with the proper tool.

TOOLS LIST

Although specialty tools aren't absolutely necessary for attaching F-connectors

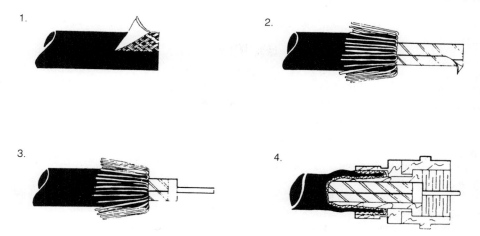

Fig. E-1. The steps in attaching an F-connector to a coaxial cable.

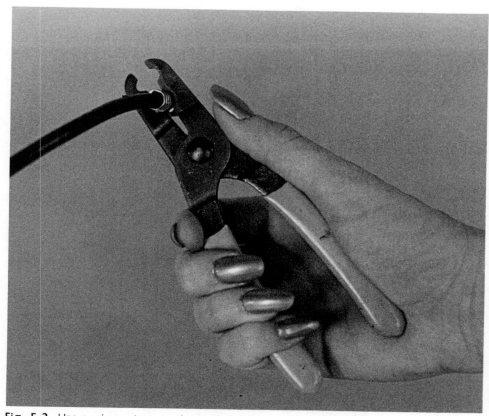

Fig. E-2. Use a crimper to securely attach the F-connector to the cable.

APPENDIX E: ATTACHING AN F-CONNECTOR

to coaxial cable, using the right tool makes the work go faster. The results are more professional, too.

You can cut the foam insulation from around the center conductor with a knife or use a conventional wire stripper. The stripper helps prevent nicking the conductor. Perhaps the best method is to use a coax cable stripper tool. This stripper, which works with all types of television coax, automatically cuts off the proper amount of outer insulation and foam.

Because F-connectors are large, they require a special tool for proper crimping (regular wire crimpers don't work well). The tool is available in a variety of styles—your choice from economy (as shown in Fig. E-2) to heavy-duty, professional models (prices are from $2 to $12).

If your cables will be used indoors, you can use crimpless F-connectors. These screw securely onto the end of the cable. Crimpless connectors are about twice as expensive as their crimped cousins, so you might not want to use them extensively.

Appendix F

Cable and
Component Signal Loss

When you send an electrical current down a wire, it gets weaker the farther it goes. If the wire is long enough, the current can be totally absorbed by the cable. With video, signal loss (attenuation) becomes apparent with cable lengths over 75 feet. Similar losses are encountered when passing a signal through a passive electrical component, such as a matching transformer or splitter.

You can counter the effects of long cable lengths and passive components in two ways: use a better quality cable or use a signal amplifier. Signal amps are ideal when you must send the signal through 100 feet or more of cable and when the signal isn't strong to begin with. But good amplifiers can be expensive.

An economical approach is to carefully choose the right cable for your needs. As a general rule, coax cable absorbs signals far more readily than twinlead—about 50 percent more, in fact. But that doesn't mean coax isn't the best choice. Coax lasts longer than twinlead and remains stable over temperature and environmental changes. Twinlead must be absolutely dry or attentuation increases 500 to 1000 percent. Loss is also greater with twinlead when the signal is split two or more ways.

It's important to note that the amount of loss depends on the frequency of the signal. The higher the frequency, the more signal will be lost. Table F-1 details the signal loss through 100 feet of RG59 and RG6 coaxial cable at different channels (higher channels are higher frequency). Table F-2 details the signal loss through common television hookup components.

APPENDIX F: CABLE AND COMPONENT SIGNAL LOSS

Table F-1. Signal Loss Through 100 Feet of Coax

Channel	RG59	RG6
2	2.6 dB	1.7 dB
6	3.5 dB	1.9 dB
7	4.9 dB	2.8 dB
13	5.4 dB	3.0 dB
20	8.3 dB	4.8 dB
40	9.2 dB	5.6 dB
60	10.3 dB	6.2 dB
83	11.9 dB	6.8 dB

Table F-2. Signal Loss Through Common Passive Components

Component	Insertion Loss	Passes Dc
300-75 ohm balun	1 dB	No
75-300 ohm balun	1 dB	No
Splitter	3 dB	Yes
75/300 ohm band splitter	1 dB	*
A-B switch	N/A	Yes

*Most band splitters pass dc current through INPUT and VHF terminals but not INPUT and UHF terminals.

Appendix G
Television
Frequency Spectrum

All television broadcast channels fall within 54 MHz to 890 MHz. This comprises all vhf low-band, vhf high-band, uhf (standard), uhf (translator), RCI sub-channels, and mid-band and super-band cable television channels.

Table G-1 shows a frequency spectrum chart that details the allocation of television channels with adjacent service frequencies, including short wave, FM radio, and ham radio. Table G-2 lists over-the-air TV channel assignments with audio, video and color-carrier frequencies. Note that the exact frequencies of the picture, color, and sound carriers might be different in cable systems using IRC and HRC frequency-offset broadcasting standards.

APPENDIX G: TELEVISION FREQUENCY SPECTRUM

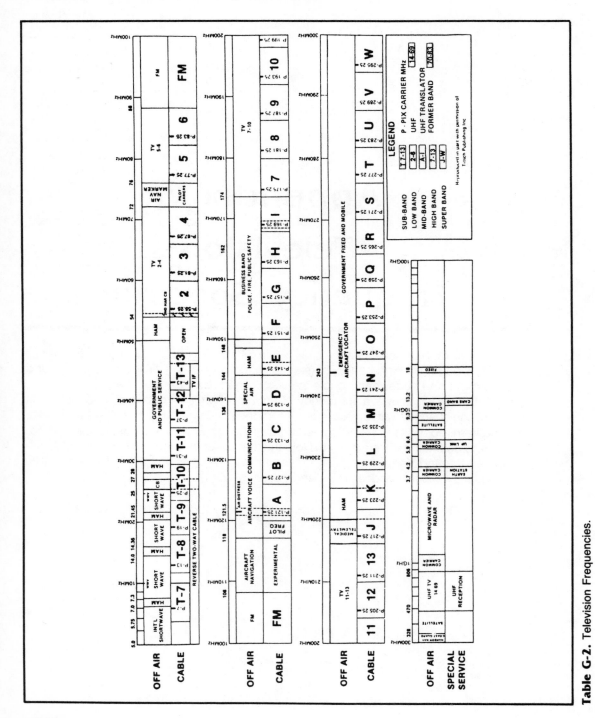

Table G-2. Television Frequencies.

Table G-2. Television Frequencies.

Channel No.	Frequency Range (MHz)	Picture Carrier (MHz)	Color Subcarrier (MHz)	Sound Carrier (MHz)
VHF LOW BAND CHANNELS				
2	54-60	55.25	58.83	59.75
3	60-66	61.25	64.83	65.75
4	66-72	67.25	70.83	71.75
5	76-82	77.25	80.83	81.75
6	82-88	83.25	86.83	87.75
FM	88-108	—	—	—
VHF HIGH BAND CHANNELS				
7	174-180	175.25	178.83	179.75
8	180-186	181.25	184.83	185.75
9	186-192	187.25	190.83	191.75
10	192-198	193.25	196.83	197.75
11	198-204	199.25	202.83	203.75
12	204-210	205.25	208.83	209.75
13	210-216	211.25	214.83	215.75
UHF CHANNELS				
14	470-476	471.25	474.83	475.75
15	476-482	477.25	480.83	481.75
16	482-488	483.25	486.83	487.75
17	488-494	489.25	492.83	493.75
18	494-500	495.25	498.83	499.75
19	500-506	501.25	504.83	505.75
20	506-512	507.25	510.83	511.75
21	512-518	513.25	516.83	517.75
22	518-524	519.25	522.83	523.75
23	524-530	525.25	528.83	529.75
24	530-536	531.25	534.83	535.75
25	536-542	537.25	540.83	541.75
26	542-548	543.25	546.83	547.75
27	548-554	549.25	552.83	553.75
28	554-560	555.25	558.83	559.75
29	560-566	561.25	564.83	565.75
30	566-572	567.25	570.83	571.75
31	572-578	573.25	576.83	577.75
32	578-584	579.25	582.83	583.75
33	584-590	585.25	588.83	589.75
34	590-596	591.25	594.83	595.75
35	596-602	597.25	600.83	601.75
36	602-608	603.25	606.83	607.75
37	608-614	609.25	612.83	613.75
38	614-620	615.25	618.83	619.75
39	620-626	621.25	624.83	625.75
40	626-632	627.25	630.83	631.75
41	632-638	633.25	636.83	637.75
42	638-644	639.25	642.83	643.75
43	644-650	645.25	648.83	649.75
44	650-656	651.25	654.83	655.75

APPENDIX G: TELEVISION FREQUENCY SPECTRUM

Channel No.	Frequency Range (MHz)	Picture Carrier (MHz)	Color Subcarrier (MHz)	Sound Carrier (MHz)
45	656-662	657.25	660.83	661.75
46	662-668	663.25	666.83	667.75
47	668-674	669.25	672.83	673.75
48	674-680	675.25	678.83	679.75
49	680-686	681.25	684.83	685.75
50	686-692	687.25	690.83	691.75
51	692-698	693.25	696.83	697.75
52	698-704	699.25	702.83	703.75
53	704-710	705.25	708.83	709.75
UHF CHANNELS				
54	710-716	711.25	714.83	715.75
55	716-722	717.25	720.83	721.75
56	722-728	723.25	726.83	727.75
57	728-734	729.25	732.83	733.75
58	734-740	735.25	738.83	739.75
59	740-746	741.25	744.83	745.75
60	746-752	747.25	750.83	751.75
61	752-758	753.25	756.83	757.75
62	758-764	759.25	762.83	763.75
63	764-770	765.25	768.83	769.75
64	770-776	771.25	774.83	775.75
65	776-782	777.25	780.83	781.75
66	782-788	783.25	786.83	787.75
67	788-794	789.25	792.83	793.75
68	794-800	795.25	798.83	799.75
69	800-806	801.25	804.83	805.75
70	806-812	807.25	810.83	811.75
UHF TRANSLATOR CHANNELS				
71	812-818	813.25	816.83	817.75
72	818-824	819.25	822.83	823.75
73	824-830	825.25	828.83	829.75
74	830-836	831.25	834.83	835.75
75	836-842	837.25	840.83	841.75
76	842-848	843.25	846.83	847.75
77	848-854	849.25	852.83	853.75
78	854-860	855.25	858.83	859.75
79	860-866	861.25	864.83	865.75
80	866-872	867.25	870.83	871.75
81	872-878	873.25	876.83	877.75
82	878-884	879.25	882.83	883.75
83	884-890	885.25	888.83	889.75
RCI SUB CHANNELS				
01	11-17	15.75	12.17	11.25
02	17-23	21.75	18.17	17.25
03	23-29	27.75	24.17	23.25
04	29-35	33.75	30.17	29.25
05	35-41	39.75	36.17	35.25

Channel No.	Frequency Range (MHz)	Picture Carrier (MHz)	Color Subcarrier (MHz)	Sound Carrier (MHz)
CATV MID AND SUPERBAND CHANNELS				
MID BAND CHANNELS				
A	120-126	121.25	124.83	125.75
B	126-132	127.25	130.83	131.75
C	132-138	133.25	136.83	137.75
D	138-144	139.25	142.83	143.75
E	144-150	145.25	148.83	149.75
F	150-156	151.25	154.83	155.75
G	156-162	157.25	160.83	161.75
H	162-168	163.25	166.83	167.75
I	168-174	169.25	172.83	173.75
SUPERBAND CHANNELS				
J	216-222	217.25	220.83	221.75
K	222-228	223.25	226.83	227.75
L	228-234	229.25	232.83	233.75
M	234-240	235.25	238.83	239.75
N	240-246	241.25	244.83	245.75
Q	246-252	247.25	250.83	251.75
P	252-258	253.25	256.83	257.75
Q	258-264	259.25	262.83	263.75
R	264-270	265.25	268.83	269.75
S	270-276	271.25	274.83	275.75
T	276-282	277.25	280.83	281.75
U	282-288	283.25	286.83	287.75
V	288-294	289.25	292.83	293.75
W	294-300	295.25	298.83	299.75

Appendix H

VCR Specifications Chart

There are hundreds of models of VCRs to choose from. This variety makes comparing features and specifications of even a small handful of decks a herculean effort. The cross-reference charts on the following pages list over 300 early- and late-model VCRs. You can use the chart when evaluating a new or used deck for purchase or to check the specifications of a model you already own or are servicing.

Brand	Model	Format	Mono/Stereo	Hi-Fi	Dig. Effects	Super	HQ	# Video Heads	Day/Event	Channels	ONP	Remote	MTS	Slo-Mo	Freeze	Fr. Adv.	PIP
Akai	VS-109U	VHS	Mono				•	2	14/1	82	•	•			•		
Akai	VS-110U	VHS	Mono					2	14/2	82	•	•					
Akai	VS-111U	VHS	Mono				•	2	14/2	107	•	•					
Akai	VS-115U	VHS	Mono				•	2	14/4	107		•			•		
Akai	VS-125U	VHS	Mono				•	2	14/8	107		•	•		•		
Akai	VS-515U	VHS	Stereo	•			•	2	14/6	107		•	•	•	•	•	
Akai	VS-525U	VHS	Stereo	•				4	14/6	107		•	•	•	•	•	
Akai	VS-565U	VHS	Stereo	•				4	14/6	107		•	•	•	•	•	
Akai	VS-626U	VHS	Stereo	•				4	28/8	142		•	•	•	•	•	•
Akai	VS-M930U	VHS	Stereo	•		•		4	n/a	n/a		•	•	•	•	•	•
Canon	VR-E10	8mm	Mono				•	2	14/4	105		•			•	•	
Canon	VR-HF600	VHS	Stereo	•				4	14/4	107		•	•	•	•		
Canon	VR-HF720	VHS	Stereo	•		•		4	30/8	107	•	•	•	•	•		
Canon	VR-HF800	VHS	Stereo					4	n/a	n/a	•	•	•	•	•		
Curtis Mathes	AV 725	VHS	Mono				•	2	68/14	68							
Curtis Mathes	AV 730	VHS	Mono				•	2	14/2	98		•					
Curtis Mathes	AV 745	VHS	Mono				•	2	14/4	98		•		•	•		
Curtis Mathes	AV 755	VHS	Stereo	•			•	2	14/4	98		•	•	•	•		
Curtis Mathes	AV 758	VHS	Stereo	•				4	21/8	98		•	•	•	•	•	
Curtis Mathes	BV 747	VHS	Stereo	•				4	n/a	n/a		•	•	•	•	•	
Curtis Mathes	BV 780	VHS	Stereo	•		•		4	n/a	n/a		•	•	•	•	•	•
Daewoo	VCR-30DAU	VHS	Mono					2	n/a	n/a							
Daewoo	VCR-50DAU	VHS	Mono					2	7/4	82							
Daewoo	DVR-8450	Beta	Mono				•	2	14/8	105		•			•		
Daewoo	DVR-8750	Beta	Mono				•	2	14/8	82							
Denon	VA-1000	VHS	Stereo	•				4	14/8	82		•	•	•	•	•	
Eastman Kodak	MVS-3360	8mm	Mono					2	14/4	133		•		•	•	•	
Eastman Kodak	MVS-5380	8mm	PCM Stereo					2	21/8	169		•		•	•	•	
Emerson	VCR872	VHS	Mono				•	2	21/8	169					•		
Emerson	VCR920	VHS	Mono				•	2	14/8	110		•			•		
Emerson	VCR951H	VHS	Mono				•	4	14/8	107		•		•	•		
Emerson	VCS955H	VHS	Stereo	•			•	4	14/4	105		•	•	•	•		
Emerson	VCS966H	VHS	Stereo	•				4	14/4	105		•	•	•	•	•	
Emerson	VCS977	VHS	Stereo	•				4	14/4	139		•	•	•	•	•	
Fisher	FVH-820	VHS	Stereo	•				2	14/5	111		•	•	•	•		
Fisher	FVH-830	VHS	Stereo	•				4	14/9	111		•	•	•	•	•	
Fisher	FVH-840	VHS	Mono				•	4	14/9	111		•		•	•		
Fisher	FVH-905	VHS	Mono				•	2	14/4	111		•		•	•		
Fisher	FVH-920	VHS	Stereo	•			•	4	14/6	111		•	•	•	•		
Fisher	FVH-940	VHS	Stereo	•				4	14/6	111	•	•	•	•	•	•	
Fisher	FVH-950	VHS	Stereo	•				4	year/8	111	•	•	•	•	•	•	
Fisher	FVH-960	VHS	Stereo	•				4	14/9	140	•	•	•	•	•	•	
Fisher	FVH-980	VHS	Stereo					4	year/8	140		•	•	•	•	•	
Fisher	FVH-990	VHS	Stereo					4	year/8	181		•	•	•	•	•	•
General Electric	9-7115	VHS	Mono				•	2	14/4	93		•					
General Electric	9-7120	VHS	Mono				•	2	14/4	93		•			•		
General Electric	9-7145	VHS	Mono				•	2	14/4	93		•		•	•		
General Electric	9-7176	VHS	Mono				•	4	14/4	93		•		•	•		
General Electric	9-7215	VHS	Stereo	•			•	4	14/4	93		•	•	•	•		
General Electric	9-7245	VHS	Mono				•	4	14/4	93		•		•	•		
General Electric	9-7250	VHS	Mono				•	4	14/4	93		•		•	•		
General Electric	9-7256	VHS	Mono				•	4	14/4	93		•		•	•		
General Electric	9-7276	VHS	Stereo	•			•	4	14/4	93		•	•	•	•	•	
General Electric	9-7320	VHS	Stereo	•			•	4	14/4	93	•	•	•	•	•		
General Electric	9-7350	VHS	Stereo	•				4	14/4	93		•	•	•	•	•	
General Electric	9-7400	VHS	Stereo	•			•	4	21/8	93		•	•	•	•	•	
Goldstar	GVH-58FM	VHS	Stereo	•			•	4	14/8	110		•		•	•		
Goldstar	GVH-1233M	VHS	Mono				•	2	14/4	110		•		•	•		

317

Brand	Model	Format	Mono/Stereo	Hi-Fi	Dig. Effects	Super	HQ	# Video Heads	Day/Event	Channels	ONP	Remote	MTS	Slo-Mo	Freeze	Fr. Adv.	PiP
Goldstar	GVH-1240M	VHS	Mono					2	14/8	110							
Goldstar	GVH-1243M	VHS	Mono				•	2	14/4	110		•			•		
Goldstar	GVH-1248M	VHS	Mono				•	2	14/4	105		•		•	•	•	
Goldstar	GVH-1400M	VHS	Mono				•	4	14/4	105		•		•	•	•	
Goldstar	GVH-8200M	VHS	Stereo	•			•	2	14/4	110		•	•		•		
Goldstar	GVH-8220M	VHS	Stereo	•			•	2	14/4	105		•	•	•	•	•	
Harman Kardon	VCD-1000	VHS	Stereo	•				2	14/4	105		•	•	•	•	•	
Harman Kardon	VCD-2000	VHS	Stereo	•				2	21/4	110		•	•	•	•	•	
Harman Kardon	VCD-4000	VHS	Stereo	•				4	21/8	110		•	•		•	•	
Hitachi	VT-1100AR	VHS	Mono				•	2	14/2	107	•	•			•		
Hitachi	VT-1310A	VHS	Mono				•	3	14/4	107		•		•	•		
Hitachi	VT-1370A	VHS	Mono		•			3	14/4	107	•	•		•	•		
Hitachi	VT-1410A	VHS	Mono				•	2	14/4	107	•	•	•	•	•		
Hitachi	VT-1570A	VHS	Stereo		•			3	14/4	107		•		•	•		
Hitachi	VT-1700A	VHS	Stereo	•			•	2	14/4	107		•			•		
Hitachi	VT-1710A	VHS	Stereo	•			•	3	14/4	119		•		•	•		
Hitachi	VT-1720A	VHS	Stereo	•			•	5	365/8	169		•		•	•		
Hitachi	VT-1800A	VHS	Stereo	•			•	4	365/8	167		•		•	•		
Hitachi	VT-2700A	VHS	Stereo	•			•	2	14/4	107	•	•		•	•		
Hitachi	VT-86A	VHS	Stereo	•				2	14/1	107		•		•	•		
Hitachi	VT-87A	VHS	Stereo	•				5	365/8	105	•	•		•	•		
Ikko	VR-8000	VHS	Mono					2	14/1	107		•		•	•	•	
Instant Replay	610 CM	VHS	Mono				•	2	14/4	104					•		
Instant Replay	610 IT3	VHS	Stereo	•			•	4	30/8	107		•		•	•	•	
Instant Replay	615 IT3	VHS	Stereo	•			•	2	30/8	107		•		•	•		
Instant Replay	618 IT3	VHS	Stereo	•			•	4	30/8	107		•		•	•	•	
Instant Replay	618 RGB	VHS	Stereo	•			•	2	14/2	107		•	•	•	•		
J C Penney	6069	VHS	Mono				•	2	14/4	99		•			•		
J C Penney	6071	VHS	Mono				•	3	14/4	107		•		•	•		
J C Penney	6072	VHS	Mono				•	4	14/4	99		•	•	•	•		
J C Penney	6073	VHS	Stereo	•			•	2	14/4	110		•	•	•	•		
J C Penney	6076	VHS	Stereo	•		•	•	4	14/4	111		•	•	•	•		
JVC	HR-D170	VHS	Mono				•	2	14/4	111		•		•	•		
JVC	HR-D180	VHS	Mono				•	2	14/4	111		•		•	•		
JVC	HR-D370	VHS	Stereo	•			•	2	14/8	181		•	•	•	•		
JVC	HR-565U	VHS	Stereo	•			•	2	14/8	181		•	•	•	•		
JVC	HR-D566	VHS	Stereo	•			•	4	14/2	107		•	•	•	•		
JVC	HR-D570	VHS	Stereo	•			•	4	n/a	n/a		•	•	•	•		
JVC	HR-D725U	VHS	Stereo	•			•	4	14/8	139		•	•	•	•		
JVC	S5000U	VHS	Stereo	•		•	•	4	14/8	181		•	•	•	•		
Kenwood	KV-917F	VHS	Mono				•	2	14/1	181		•		•	•		
Lloyd's	L824	VHS	Mono				•	2	14/4	110		•			•		
Lloyd's	L828	VHS	Mono				•	2	21/4	110					•		
Lloyd's	L840	VHS	Mono				•	2	14/2	110		•			•		
Magnavox	VR9510AT	VHS	Mono				•	2	14/4								
Magnavox	VR9520AT	VHS	Mono				•	2	14/4	93		•			•		
Magnavox	VR9530AT	VHS	Mono				•	2	14/4	93		•		•	•		
Magnavox	VR9540AT	VHS	Stereo	•			•	4	14/4	93		•	•	•	•		
Magnavox	VR9550AT	VHS	Mono				•	4	111/4	93		•		•	•		
Magnavox	VR9558AT	VHS	Stereo	•			•	4	14/8	93		•	•	•	•		
Magnavox	VR9560AT	VHS	Stereo	•			•	4	21/8	155	•	•	•	•	•		
Magnavox	VR9565AT	VHS	Stereo	•			•	2	21/8	155	•	•	•	•	•		
Magnavox	VR9642	VHS	Mono				•	4	21/8	155	•	•		•	•		
Magnavox	VR9652	VHS	Stereo	•			•	4	21/8	155	•	•	•	•	•		
Magnavox	VR9665	VHS	Stereo	•		•	•	4	21/8	155	•	•	•	•	•		•
Magnavox	VR9670	VHS	Stereo	•			•	4	21/4	110		•	•	•	•		
Marantz	VR460HQ	VHS	Stereo	•			•	4	21/4	110		•	•	•	•		
Marantz	VR465HQ	VHS	Stereo	•			•	4	21/8	140		•	•	•	•		
Marantz	VR550HQ	VHS	Stereo	•			•	4	21/8			•	•	•	•		

Brand	Model	Format	Mono/Stereo	Hi-Fi	Dig. Effects	Super	HQ	# Video Heads	Day/Event	Channels	OMP	Remote	MTS	Slo-Mo	Freeze	Fr. Adv.	PIP
Marantz	VR560HQ	VHS	Stereo	•			•	4	21/8	140							
Minolta	MV-205	VHS	Mono					2	14/4	107		•			•		
Minolta	MV-605	VHS	Stereo	•				3	14/4	119		•	•		•	•	
Mitsubishi	HS-337UR	VHS	Mono				•	2	14/4	107		•		•	•	•	•
Mitsubishi	HS-339UR	VHS	Mono				•	4	14/8	107		•		•	•	•	
Mitsubishi	HS411UR	VHS	Stereo	•				4	14/8	107		•		•	•	•	
Mitsubishi	HS421UR	VHS	Stereo	•	•			4	14/8	139	•	•		•	•	•	•
NEC	DX-1000U	VHS	Mono				•	2	21/4	110	•	•	•	•	•		
NEC	DX-2000U	VHS	Stereo	•	•			2	21/4	110	•	•	•	•	•	•	
NEC	DX-5000U	VHS	Stereo	•	•		•	2	21/4	110	•	•	•	•	•	•	
NEC	N-915U	VHS	Mono				•	2	21/4	110		•		•	•		
NEC	N-925U	VHS	Mono				•	4	21/4	110		•		•	•	•	
NEC	N-945U	VHS	Stereo	•			•	2	21/4	110		•	•	•	•		
NEC	N-955U	VHS	Stereo	•			•	2	21/4	110	•	•	•	•	•		
NEC	N-965U	VHS	Stereo	•			•	4	21/8	140		•	•	•	•	•	
NEC	N-951U	VHS	Stereo	•			•	2	21/4	140		•	•	•	•		
NEC	N-961U	VHS	Stereo	•			•	4	--	140		•	•	•	•	•	
Panasonic	PV-1360	VHS	Mono				•	2	14/2	68					•		
Panasonic	PV-1361	VHS	Mono				•	2	14/2	93		•		•	•		
Panasonic	PV-1461	VHS	Stereo	•			•	2	14/4	99		•	•	•	•		
Panasonic	PV-1462	VHS	Stereo	•			•	2	14/4	99		•	•	•	•		
Panasonic	PV-1464	VHS	Mono				•	2	14/4	99		•			•		
Panasonic	PV-1560	VHS	Mono				•	4	14/4	99		•		•	•	•	
Panasonic	PV-1563	VHS	Stereo	•			•	4	14/4	99		•	•	•	•	•	
Panasonic	PV-1564	VHS	Stereo	•			•	4	14/4	99		•	•	•	•	•	
Panasonic	PV-1565	VHS	Mono				•	4	14/4	99		•		•	•	•	
Panasonic	PV-1642	VHS	Stereo	•			•	4	14/4	99		•	•	•	•	•	
Panasonic	PV-1740	VHS	Stereo	•			•	4	21/8	169		•	•	•	•	•	
Panasonic	PV-1742	VHS	Stereo	•	•		•	4	21/8	155		•	•	•	•	•	•
Panasonic	PV-3770	VHS	Stereo	•	•			3	n/a	n/a		•	•	•	•	•	•
Panasonic	PV-4770	VHS	Stereo	•		•		4	14/4	n/a		•	•	•	•		
Panasonic	PV-4780	VHS	Stereo	•		•		4	n/a	n/a		•	•	•	•	•	•
Panasonic	PV-9600A	VHS	Stereo	•				4	14/8	139		•		•	•		
Panasonic	PV-S4764	VHS	Stereo	•				4	30/8	15		•		•	•		
Pentax	PV T100A	VHS	Stereo	•				2	14/4	107		•	•	•	•		
Pentax	PV T150A	VHS	Mono					3	14/4	119		•		•	•		
Philco	VJ8760AT	VHS	Mono					2	14/4	93		•		•	•		
Philco	VJ8765AT	VHS	Stereo	•				2	14/4	93		•	•	•	•		
Philco	VJ8770AT	VHS	Mono					4	14/4	93		•		•	•		
Pioneer	VH-600	VHS	Stereo	•				2	14/4	107		•	•	•	•		
Pioneer	VH-900	VHS	Stereo	•				5	14/4	119		•	•	•	•	•	•
Pioneer	VX-90	Beta	Stereo	•					21/8	181		•	•	•	•		
Quasar	VH-5165	VHS	Mono					2	14/4	107		•		•	•		
Quasar	VH-5168	VHS	Stereo	•				4	14/4	107		•	•	•	•		
Quasar	VH-5260	VHS	Mono					4	14/4	93	•	•		•	•		
Quasar	VH-5261	VHS	Mono					2	14/4	93		•		•	•		
Quasar	VH-5262	VHS	Mono					4	14/4	93		•		•	•		
Quasar	VH-5268	VHS	Stereo	•				4	14/4	93		•	•	•	•		
Quasar	VH-5379	VHS	Stereo	•				4	14/4	107		•	•	•	•		
Quasar	VH-5479	VHS	Mono					3	n/a	n/a		•		•	•	•	•
Quasar	VH-5665	VHS	Stereo	•				4	21/8	107		•	•	•	•		
Quasar	VH-5865	VHS	Stereo	•				4	14/4	155		•	•	•	•	•	
Quasar	VH-5975	VHS	Stereo	•			•	4	21/8	n/a	•	•	•	•	•	•	
RCA	VLT700HF	VHS	Mono				•	5	365/8	133		•		•	•	•	
RCA	VMT285	VHS	Mono				•	2	14/2	107		•		•	•		
RCA	VMT385	VHS	Mono				•	3	14/4	107		•		•	•		
RCA	VMT390	VHS	Mono					3	n/a	n/a		•			•		
RCA	VMT400	VHS	Mono				•	3	365/4	119		•		•	•	•	
RCA	VMT495	VHS	Stereo	•			•	3	365/4	119		•	•	•	•		•

319

Brand	Model	Format	Mono/Stereo	Hi-Fi	Dig. Effects	Super	HQ	# Video Heads	Day/Event	Channels	ONP	Remote	MTS	Slo-Mo	Freeze	Fr. Adv.	PIP
RCA	VMT590	VHS	Mono				•	5	365/4	119	•	•	•	•	•	•	
RCA	VMT595	VHS	Stereo					5	365/4	119	•	•	•	•	•	•	
RCA	VMT630HF	VHS	Stereo	•			•	3	365/4	119	•	•	•	•	•	•	
RCA	VMT670HF	VHS	Stereo	•			•	5	365/8	169	•	•	•	•	•	•	
RCA	VPT640	VHS	Stereo			•		4	365/8	169		•		•	•		
Radio Shack	Model 17	VHS	Mono					4	14/5	111		•			•		
Radio Shack	Model 18	VHS	Mono					2	14/5	105					•		
Radio Shack	Model 22	Beta	Stereo	•			::	2	14/8	105		•		•	•		
Radio Shack	Model 41	VHS	Stereo	•				2	14/5	110		•			•		
STS	VR400	VHS	Stereo	•				2	14/3	82		•			•	•	
Samsung	VR310	VHS	Mono					2	14/6	105		•			•		
Samsung	VR550H	VHS	Stereo	•				2	14/6	105		•	•	•	•		
Samsung	VT226T	VHS	Stereo					4	14/6	105	•	•	•		•		
Samsung	VT290T	VHS	Stereo					4	28/8	139					•		
Samsung	VT311T	VHS	Mono					2	14/6	82		•			•		
Samsung	VR6600F	VHS	Stereo				•	2	14/6	82	•	•			•	•	
Sansui	SV-R5700	VHS	Mono				•	4	14/4	108		•		•	•	•	
Sansui	SV-R7700	VHS	Mono					2	14/4	108		•			•	•	
Sansui	SV-R9500HF	VHS	Stereo	•			•	4	14/4	178	•	•		•	•	•	
Sansui	SV-R9700HF	VHS	Stereo	•			•	4	14/8	178		•		•	•	•	
Sanyo	VCR 4027	Beta	Mono				::	2	7/1	105					•		
Sanyo	VCR 4030	Beta	Mono				::	2	14/8	105		•			•		
Sanyo	VCR 7150	Beta	Stereo				::	2	14/8	105		•			•		
Sanyo	VCR 7250	Beta	Stereo				::	2	14/8	105		•		•	•		
Sanyo	VCR 7500	Beta	Stereo					3	14/4	107		•		•	•		
Sanyo	VHR 1600	VHS	Mono	•			•	4	14/8	107		•		•	•		
Sanyo	VHR 1900	VHS	Stereo	•			•	2	365/8	111	•	•	•	•	•		
Sanyo	VHR 2900	VHS	Stereo				•	2	14/8	n/a		•			•		
Scott	SVR-110	VHS	Mono				•	2	14/4	n/a		•			•	•	
Scott	SVR-330S	VHS	Stereo				•	4	14/8	139		•			•		
Sears	SVR-504S	VHS	Stereo					2	14/4	82		•			•		
Sears	5329	VHS	Mono					2	14/4	107					•		
Sears	5331/331	VHS	Stereo				•	4	14/4	107		•	•		•		
Sears	5342	Beta	Mono					2	3/1	82					•		
Sears	53091	Beta	Mono			•		2	n/a	n/a					•		
Sharp	VC-D75U	VHS	Stereo	•			•	4	14/5	108	•	•	•	•	•	•	
Sharp	VC-5F7U	VHS	Mono	•			•	2	14/4	110		•		•	•		
Sharp	VC-685U	VHS	Mono	•			•	2	14/4	110		•		•	•		
Sharp	VC-686U	VHS	Stereo				•	4	14/4	110	•	•	•	•	•	•	
Sharp	VC-687U	VHS	Stereo				•	2	14/4	110		•		•	•	•	
Sharp	VC-H64U	VHS	Stereo	•				2	14/4	110		•			•		
Sharp	VC-H65U	VHS	Stereo	•			•	2	14/4	110		•			•		
Sharp	VC-6846U	VHS	Mono				•	2	14/4	110	•	•			•		
Sharp	VC-6847U	VHS	Mono				•	4	14/4	110	•	•			•		
Sharp	VC-686U	VHS	Mono				•	2	14/4	110	•	•			•		
Sharp	VC-787U	VHS	Mono				•	4	14/6	110	•	•			•		
Sharp	VC-799U	VHS	Mono				•	4	14/6	110		•			•	•	
Sharp	VC-7854U	VHS	Stereo	•			•	2	14/4	110		•	•	•	•	•	
Sharp	VC-7857U	VHS	Stereo	•			•	2	14/4	110		•	•	•	•	•	
Sharp	VC-7980	VHS	Stereo				•	2	14/4	110		•			•		
Sharp	VC-T64U	VHS	Mono				•	2	14/5	110		•			•	•	
Sony	EV-C8U	8mm	PCM Stereo				::	2	--	--					•	•	
Sony	EV-S700U	8mm	Mono				::	2	7/3	148		•		•	•	•	
Sony	EV-A80	8mm	Mono				::	2	7/6	148		•		•	•	•	
Sony	SL100	Beta	Stereo				::	2	7/6	148		•		•	•	•	
Sony	SL350	Beta	Mono			•	::	2	7/6	148		•		•	•	•	
Sony	SL700	Beta	Mono				::	2	21/4	107		•		•	•	•	
Sony	SL2410	Beta	Mono				::	2	7/6	107		•		•	•	•	
Sony	SL-HF300	Beta	Stereo	•			::	2	7/6	107		•		•	•	•	

Brand	Model	Format	Mono/Stereo	Hi-Fi	Dig. Effects	Super	HQ	# Video Heads	Day/Event	Channels	OnP	Remote	MTS	Slo-Mo	Freeze	Fr. Adv.	PIP
Sony	SL-HF400	Beta	Stereo	•			..	2	7/6	169		•	•	•	•	•	
Sony	SL-HF450	Beta	Stereo	•			..	2	7/6	148		•	•	•	•	•	
Sony	SL-HF550	Beta	Stereo	•		•	..	3	7/6	??		•	•	•	•	•	
Sony	SL-HF600	Beta	Stereo	•		•	..	3	7/6	181		•	•	•	•	•	
Sony	SL-HF750	Beta	Stereo	•		•	..	4	21/6	181		•	•	•	•	•	
Sony	SL-HF900	Beta	Stereo	•		•	..	4	21/8	181		•	•	•	•	•	
Sony	SL-HF1000	Beta	Stereo	•		•	..	4	21/8	181		•	•	•	•	•	
Sony	SL-HFR30	Beta	Mono				..	2	7/6	105		•		•	•	•	
Sony	SL-HFR70	Beta	Mono				..	2	7/6	148		•		•	•	•	
Supra	SV-1800	VHS	Mono					2	14/4	83					•		
Supra	SV-1900	VHS	Mono	•				2	14/4	83				•	•		
Sylvania	VC4546SL	VHS	Stereo	•			•	4	14/8	139		•	•	•	•	•	
Sylvania	VC8930	VHS	Mono				•	2	14/4	93		•		•			
Sylvania	VC8940	VHS	Mono				•	2	14/4	93		•		•	•		
Sylvania	VC8945	VHS	Mono				•	2	14/4	93		•		•	•	•	
Sylvania	VC8950	VHS	Mono				•	4	14/4	93		•		•	•		
Sylvania	VC8960	VHS	Mono				•	4	14/4	93		•		•	•	•	
Sylvania	VC9870	VHS	Stereo	•			•	4	21/8	93		•	•	•	•	•	
Tatung	VRH-8500U	VHS	Mono				•	4	14/4	105		•		•	•	•	
Tatung	VRH-8700U	VHS	Mono				•	4	14/4	108		•		•	•	•	
Tatung	VRH-8800U	VHS	Mono				•	4	14/4	105		•		•	•	•	
Teac	MV 350	VHS	Stereo	•			•	2	14/4	110		•		•			
Teac	MV 600	VHS	Stereo	•			•	2	14/4	110		•		•	•		
Teac	MV 1000	VHS	Mono	•		•	•	2	14/8	110		•		•	•	•	
Teknika	VCR663	VHS	Mono				•	2	14/4	93		•		•			
Teknika	VCR684	VHS	Mono				•	2	14/4	93		•		•	•		
Teknika	VCR689	VHS	Stereo	•			•	4	14/4	93		•	•	•	•		
Toshiba	DX-3	VHS	Mono				..	2	14/4	117		•		•	•		
Toshiba	DX-7	VHS	Stereo	•		•		4	7/4	117		•	•	•	•		
Toshiba	DX-400	VHS	Stereo	•				2	n/a	n/a		•	•				
Toshiba	DX-800	VHS	Stereo	•		•		4	n/a	n/a		•	•	•	•		
Toshiba	M-2200	VHS	Mono					4	n/a	n/a		•		•	•		
Toshiba	M-2400	VHS	Stereo	•				2	7/4	117		•	•	•	•		
Toshiba	M-2700	VHS	Mono					2	7/4	117		•		•	•		
Toshiba	M-4100	VHS	Stereo	•				4	7/4	117		•	•	•	•		
Toshiba	M-4200	VHS	Mono					4	7/4	117		•		•	•		
Toshiba	M-4500	VHS	Stereo	•				4	7/4	117		•	•	•	•		
Toshiba	M-5530	VHS	Mono					4	7/4	117		•		•	•		
Toshiba	M-5900	VHS	Stereo	•				2	7/1	105		•	•	•	•		
Vector Research	V-M50	Beta	Mono				..	2	7/4	105		•		•	•		
Vector Research	V-2020	VHS	Mono				•	2	14/6	110		•		•	•		
Vector Research	V-4020	VHS	Stereo	•			•	2	21/4	110		•	•	•	•		
Vector Research	V-4040	VHS	Stereo	•			•	4	21/4	110		•	•	•	•		
Vector Research	V-5040	VHS	Stereo	•			•	4	21/8	140		•	•	•	•	•	
Yamaha	YV-700	VHS	Stereo	•			•	4	14/4	181		•	•	•	•	•	
Yamaha	YV-1000	VHS	Stereo	•			•	4	14/8	181		•	•	•	•	•	
Zenith	VR1820	VHS	Mono				•	4	14/4	178	•	•		•	•		
Zenith	VR1825	VHS	Stereo	•			•	4	14/4	178	•	•	•	•	•		
Zenith	VR1830	VHS	Mono				•	4	14/4	178	•	•		•	•		
Zenith	VR1870	VHS	Mono				•	4	14/4	178	•	•		•	•		
Zenith	VR2220	VHS	Stereo	•			•	4	14/4	178	•	•	•	•	•	•	
Zenith	VR2300	VHS	Stereo	•			•	4	14/4	178	•	•	•	•	•	•	
Zenith	VR3220	VHS	Stereo	•			•	4	14/4	178	•	•	•	•	•	•	
Zenith	VR3300	VHS	Stereo	•			•	4	14/8	178	•	•	•	•	•	•	
Zenith	VR4100	VHS	Mono				•	4	14/8	178	•	•		•	•	•	
Zenith	VR5100	VHS	Stereo	•			•	4	14/8	178	•	•	•	•	•	•	
Zenith	VRD700HF	VHS	Stereo	•			•	4	n/a	n/a	•	•		•	•	•	

APPENDIX H: VCR SPECIFICATIONS CHART

CAMCORDERS

Brand	Model	Format	Tape Speeds	Max Rec. Time	Play In Cam.	HQ	Flying Erase	Pickup Type	Pickup Size	V'finder Type	Auto Focus	Min. Illum.	Lens Aperture	Zoom Ratio	Power Zoom	Weight	Dimensions
Aiwa	CV-80	8mm	SP	120 min.	•			CCD	1/2"	Elec.		19 lux	f/1.4	6:1	•	5.1 lbs.	7 x 5 x 15
Canon	VM-E2	8mm	SP	120 min.	•		•	CCD	1/2"	Elec.	•	8 lux	f/1.2	6:1	•	3.5 lbs.	6 x 5.7 x 11.8
Chinon	C8-C60	8mm	SP	120 min.	•		•		1/2"	Elec.		7 lux	f/1.2	6:1	•	5.6 lbs.	15 x 8 x 5
Chinon	CV-T60	VHS	SP	160 min.	•		•	New	1/2"	Elec.		7 lux	f/1.2	6:1	•	5.6 lbs.	15 x 8 x 5
Chinon	CV-T60G	VHS	SP	160 min.	•			New	1/2"	Elec.		7 lux	f/1.2	6:1	•	5.6 lbs.	15 x 8 x 5
Curtis Mathis	AV800	VHS	SP/LP	n/a	•			n/a	?	Elec.	?	9 lux	f/1.6	6:1	•	2.8 lbs.	n/a
Fisher	FVC-801	8mm	SP	120 min.	•			New	1/2"	Elec.	?	9 lux	f/1.2	6:1	•	5.6 lbs.	9 x 5 x 15
GE	9-9606	VHS	SP	160 min.	•		•	New	1/2"	Elec.	•	7 lux	f/1.2	6:1	•	5.6 lbs.	9 x 5 x 15
GE	9-9608	VHS	SP	160 min.	•		•	New	1/2"	Elec.	•	7 lux	f/1.2	8:1	•	5.6 lbs.	9 x 5 x 15
GE	9-9610	VHS	SP	160 min.	•		•		1/2"	Elec.	•	7 lux	f/1.4	6:1	•	4.5 lbs.	5 x 6 x 14
Goldstar	GS-8A1	8mm	SP	120 min.	•			CCD	2/3"	Elec.	•	19 lux	f/1.4	6:1	•	6.5 lbs.	8 x 5 x 15
Goldstar	GVM-70AF	VHS	SP	240 min.	•		•	CCD	2/3"	Elec.	•	19 lux	f/1.2	6:1	•	6.4 lbs.	7.5 x 6.5 x 14.1
Hitachi	VM5000A	VHS	SP	160 min.	•		•	MOS	1/2"	Elec.	•	7 lux	f/1.2	6:1	?	5.5 lbs.	n/a
Hitachi	VM2100A	VHS	SP	n/a	•			Saticon	1/3"	Elec.	•	10 lux	f/1.2	6:1	•	7.5 lbs.	8 x 6 x 14
Instant Replay	6601T3	VHS	SP/LP	160 min.	•		•	CCD	?	Optical	•	20 lux	f/1	0:0	•	2.8 lbs.	4 x 4 x 1
Instant Replay	Ultra	8mm	SP	160 min.	•		•	New	1/2"	Elec.	•	7 lux	f/1.2	6:1	•	7 lbs.	9 x 15 x 5
JC Penney	686-5355	VHS	SP	160 min.	•		•		1/2"	Elec.	•	7 lux	f/1.2	6:1	•	3.9 lbs.	6 x 10 x 5
JC Penney	686-5600	VHS-C	SP	60 min.	•		•	New	1/2"	Elec.	•	15 lux	f/1.6	6:1	•	2.9 lbs.	7 x 5 x 9
JVC	GR-C7	VHS-C	SP	60 min.	•		•	CCD	1/2"	Optical	•	10 lux	f/1.6	?	?	1.7 lbs.	4 x 4 x 8
JVC	GR-C9	VHS-C	SP	60 min.	•		•		1/2"	Elec.	•	10 lux	f/1.2	6:1	•	4.2 lbs.	5 x 6 x 12
Kodak	MVS-3440	8mm	SP	240 min.	•		•	New	1/2"	Elec.	•	10 lux	f/1.4	6:1	•	4.6 lbs.	5 x 6 x 12
Kodak	MVS-3460	8mm	SP	240 min.	•		•	New	1/2"	Elec.	•	19 lux	f/1.2	6:1	•	5.1 lbs.	8 x 5 x 14
Kyocera	KD-200K	8mm	SP	120 min.	•		•	CCD	2/3"	Elec.	•	19 lux	f/1.4	6:1	•	3.1 lbs.	6 x 7 x 10
Kyocera	KD-1100U	8mm	SP	120 min.	•		•	CCD	2/3"	Elec.	•	16 lux	f/1.6	2.5:1	•	5.6 lbs.	9 x 15 x 5
Magnavox	VR8293	VHS	SP/EP	160 min.	•		•		1/2"	Elec.	•	7 lux	f/1.2	8:1	•	3.1 lbs.	6.2 x 5.2 x 10.2
Magnavox	VR8297	VHS-C	SP/EP	60 min.	•		•	New	2/3"	Elec.	•	7 lux	f/1.2	6:1	•	5.5 lbs.	7 x 8 x 14
Minolta	CR-1200SAF	VHS	SP	160 min.	•		•	MOS	2/3"	Elec.	•	7 lux	f/1.2	6:1	•	3.2 lbs.	5.4 x 4.9 x 12.7
Minolta	CR8000SAF	8mm	SP	120 min.	•		•	CCD	n/a	Elec.	•	7 lux	f/1.2	8:1	•	5.6 lbs.	15 x 9 x 5
NEC	V30U	VHS	SP	160 min.	•			New	1/2"	Elec.	•	10 lux	f/1.2	6:1	•	n/a	n/a
NEC	V40U	VHS	SP	160 min.	•		•		n/a	Elec.	•	7 lux	f/1.2	6:1	•	5.6 lbs.	4 x 8 x 13
NEC	EM8	8mm	SP	120 min.	•		•		1/2"	Elec.	•	7 lux	f/1.2	8:1	•	5.6 lbs.	14 x 4 x 8
Olympus	VX403	VHS	SP	160 min.	•		•		2/3"	Elec.	•	7 lux	f/1.2	8:1	•	5.3 lbs.	7 x 4 x 9
Olympus	VX404	8mm	SP	120 min.	•		•		2/3"	Elec.	•	7 lux	f/1.6	6:1	•	2.9 lbs.	6 x 5 x 11
Olympus	VX801	VHS	SP	120 min.	•		•		2/3"	Elec.	•	7 lux	f/1.6	6:1	•	3.1 lbs.	6 x 10 x 5
Panasonic	PV-100	VHS-C	SP	60 min.	•		•	New	n/a	Elec.	•	7 lux	f/1.2	6:1	•	5.6 lbs.	9 x 5 x 15
Panasonic	PV-220	VHS	SP	160 min.	•		•	MOS	1/2"	Elec.	•	10 lux	f/1.2	8:1	•	5.3 lbs.	7 x 5 x 14
Panasonic	PV-300	8mm	SP	160 min.	•		•		2/3"	Elec.	•	15 lux	f/1.6	6:1	•	2.9 lbs.	7 x 6 x 15
Pentax	PV-C800A	8mm	SP	120 min.	•		•	New	2/3"	Optical	•	21 lux	f/1.4	6:1	•	5.7 lbs.	7 x 6 x 15
Philco	VCR807	VHS	SP	160 min.	•		•		1/2"	Optical	•	15 lux	f/1.6	0:0	•	3.2 lbs.	4 x 4 x 9
Quasar	VM-20	8mm	SP	160 min.	•		•		2/3"	Optical	•	15 lux	f/1.6	25:1	•	5.7 lbs.	6 x 5 x 11
Quasar	VN-50	VHS-C	SP	20 min.	•		•		2/3"	Elec.	•	9 lux	f/1.4	6:1	•	3.1 lbs.	5 x 8 x 14
RCA	CPR100	VHS-C	SP	60 min.	•		•	MOS	2/3"	Elec.	•	10 lux	f/1.4	6:1	•	3 lbs.	6 x 5 x 9
RCA	CMR300	VHS	SP	160 min.	•		•	MOS	2/3"	Elec.	•	7 lux	f/1.6	6:1	•	5 lbs.	5 x 8 x 14
Sanyo	VM-8	8mm	SP	n/a	•			n/a	n/a	Elec.		19 lux	f/1.2	6:1	•	5.3 lbs.	4 x 8 x 15
Sears	53721	VHS	SP	160 min.	•		•	Saticon	1/2"	Elec.	•	7 lux	f/1.4	6:1	•	5.3 lbs.	n/a
Sharp	VC-C20UA	VHS	SP	160 min.	•		•	MOS	2/3"	Optical	•	10 lux	f/1.2	6:1	•	2.9 lbs.	n/a
Sharp	VC-C50UA	VHS-C	SP	60 min.	•		•	CCD	1/2	Optical	•	15 lux	f/1.6	0:0	•	5.5 lbs.	4 x 3 x 8
Sony	BMC-660K	Beta	SP	200 min.	•			CCD	2/3"	Optical	•	16 lux	f/1.4	6:1	•	5.7 lbs.	8.7 x 14.9 x 4.6
Sony	BMC-1000K	8mm	SP	180 min.	•			CCD	2/3"	Optical	•	21 lux	f/1.6	6:1	•	3.2 lbs.	5 x 5 x 12
Sony	CCD-M8U	8mm	SP	120 min.	•			CCD	2/3"	Optical	•	15 lux	f/1.6	0:0	•	n/a	n/a
Sony	CCD-M9U	8mm	SP	120 min.	•			CCD	2/3"	Optical	•	15 lux	f/1.6	0:0	•	5.6 lbs.	9 x 15 x 5
Sony	CCD-V3	8mm	SP	120 min.	•		•	CCD	2/3"	Optical	•	9 lux	f/1.6	25:1	•	3.1 lbs.	6 x 10 x 5
Sony	CCD-V110	8mm	SP	120 min.	•		•	CCD	2/3"	Elec.	•	14 lux	f/1.4	6:1	•	3 lbs.	6 x 5 x 9
Sony	CCD-V8AFU	8mm	SP	120 min.	•		•	CCD	2/3"	Elec.	•	6 lux	f/1.6	6:1	•	5 lbs.	5 x 8 x 14
Sylvania	VCC155	VHS-C	SP	60 min.	•		•	CCD	2/3"	Elec.	•	7 lux	f/1.6	6:1	•	5.2 lbs.	4 x 8 x 15
Sylvania	VCC157	VHS	SP	160 min.	•		•		1/2"	Elec.	•	9 lux	f/1.6	6:1	•	3.1 lbs.	n/a
Vivitar	VM-8350	8mm	SP	n/a	•			n/a	n/a	Elec.		9 lux	f/1.6	0:0	•	n/a	n/a
Zenith	VM6350	VHS-C	SP/EP	60 min.	•		•	CCD	1/2"	Optical	•	10 lux	f/1.6	0:0	•	2.8 lbs.	4 x 3 x 8
Zenith	VM6200	VHS-C	SP/EP	60 min.	•		•	CCD	1/2"	Elec.	•	15 lux	f/1.6	6:1	•	2.2 lbs.	7 x 5 x 9
Zenith	VM7100	VHS	SP/EP	480 min.	•		•	CCD	1/2"	Elec.	•	8 lux	f/1.2	6:1	•	5.3 lbs.	8 x 6 x 13

Glossary

A/B Switch — A two-way switch that is used to alternately divert a signal one of two ways. Can also be used to select from either of two incoming signals.

ac — Alternating current. Current that fluctuates to positive and negative values about a zero point. The current available at wall outlets.

afc — Automatic frequency control, a circuit built into some VCRs and TVs to automatically lock onto an incoming channel.

agc — Automatic gain control. On a TV or VCR, agc is a circuit that automatically adjusts the incoming signal to the proper levels for display or recording. On a video camera, agc is a circuit that automatically adjusts the sensitivity of the pickup tube to render the most pleasing image.

attenuator — A passive (that is, not powered by electricity) device used to reduce the power of a signal.

balance — Equal signal strength provided to both left and right stereo output channels. Balance can be adjusted to provide more signal strength to one channel than the other.

balun — See *matching transformer*.

baseband — Separate audio and video signals from a VCR or similar signal source.

Beta — A VCR format pioneered by Sony in the mid 1970's.

Beta I/II/III — The three tape record/playback speeds used in Beta decks. Beta I speed is fastest; Beta III is slowest.

bit — A binary number: 1 or 0. A certain number of bits makes a *byte*. In most computer applications, eight bits make a byte.

C-band — The frequency band used for most commercial satellite television transmission.

GLOSSARY

CATV—An abbreviated term for Community Antenna Television, now generally regarded as "cable TV."

CCD—Short for charge coupled device, a type of integrated circuit, sensitive to light, used as an imaging device in video cameras.

CCTV—Acronym for closed-circuit television. The term generally applies to "cable TV."

CRT—Short for cathode ray tube, the screen in a TV.

CTL—An acronym for "control," having two meanings: 1. The indexing/address system used on the latest models of VHS video cassette recorders. 2. An abbreviation for control, in reference to the control track on a tape or a control head in a VCR.

cable converter—An electronic device used to tune into channels broadcast through a cable system. Different from a cable decoder, which unscrambles an encrypted video/audio signal.

camcorder—A combination video camera and VCR.

capstan—A motor-driven shaft that pulls the video tape through the transport mechanism. The capstan is supplemented by the pinch roller.

carrier—1. A radio frequency signal that can have another signal superimposed over it. 2. The mechanism that holds the cassette inside the VCR; part of the cassette-life mechanism.

cassette—A cartridge that holds video tape. Also called the shell.

cassette-in switch—A small leaf switch used to detect when a cassette tape is fully loaded in the VCR.

cassette-lift mechanism—The mechanism that brings the cassette into threading position. The cassette-lift mechanism includes the carrier.

chroma—Short for *chrominance*. Chroma is the color component of the video signal.

clipping—An effect of distortion where the peaks of driven signals are chopped off. Clipping usually occurs in the amplifier when it is turned up too high, but can also occur in maladjusted circuits in a VCR or TV set.

clock—1. An electronic "metronome" used for the purpose of timing signals in a VCR's digital circuits. 2. The timing of a sequence of signals.

coax—Short for *coaxial*, a type of rounded cable made with a solid center conductor that's completely encased in a plastic covering. Around the plastic is a jacket of braided shielding wire and sometimes an aluminum sheath. The popular type of cabling used for connecting video components.

color burst—The signal, at approximately 3.57 MHz in the video bandwidth, that stores the instantaneous intensity and hue of the color for a particular spot in the TV image.

color-under—A method of placing the color (or chroma) information below the luminance information in the video bandwidth. The color-under (or heterodyne) method is used to enable accurate recording and playback of the video image.

comb filter—An electric filtering system designed to pass a certain set of frequencies but reject others.

composite video—A picture signal combined with synchronization and (possibly) color information. Usually called *baseband video*, or just *video*.

common—The ground point of a circuit or a common path for an electrical signal.

control head—The magnetic head in a VCR that records and plays back the control track.

control track—A linear track, consisting of 30- or 60-Hz pulses, placed on the bottom of video tape that aids in proper playback of the video signal.

crosstalk—1. A signal from one stereo channel that bleeds into the other. 2. A signal from a video track on a tape bleeding into the signal on the adjacent track.

cue—1. A VCR control that runs the tape in FAST-FORWARD or REWIND mode while the picture remains on the screen (same as fast-scan). 2. A signal recorded on some tapes to identify the start of a segment of video.

dc—Direct current. Current such as that from a battery where the voltage level remains the same (either positive or negative) in respect to ground.

decibel (dB)—A unit of power measurement. A 6 dB rise in signal strength represents a 100 percent increase (or doubling) in power.

demodulate—To remove the carrier signal and leave only baseband audio and video.

depth multiplex recording—The process of recording high-fidelity audio on VHS hi-fi VCRs. The system uses separate rotating heads to record the audio portion.

dew sensor—A passive electrical device used in most VCRs that detects the presence of moisture. Most dew sensors measure resistance—the resistance drops when moisture is present.

digital—A signal composed of two signal states—on or off (often expressed as "1's" and "0's"). The alternate to digital is analog, where the signal is continuously variable.

Dolby—The trade name for a popular audio noise reduction system.

down-converter—An electronic device that converts very high frequency satellite signal to a more manageable frequency, often 70 MHz or a block of frequencies between 430 to 930 MHz.

dropout—Missing information. In video, dropout is usually caused by missing oxide emulsion on the tape.

dub—To copy a tape.

dynamic range—The range of volume from softest to loudest sounds (the actual sound levels themselves are not a consideration, but it usually starts at "no sound" and goes from there). Expressed in dB (decibels). The higher the dB, the wider the dynamic range. Also used to denote picture brightness range.

EP—Extended play, the slowest speed on a VHS VCR. On earlier models, EP was called SLP, probably for "Super Long Play." See also *SP* and *LP*.

end-of-tape sensor—A sensor system used in all VCRs to detect the start or end of the video tape.

F-connector—The standard connector used with coaxial cable and the rf inputs/outputs of most video equipment.

feedhorn—In satellite television systems, a signal "collector" placed in front of a low noise amplifier.

field—One half of a video field, comprising the odd or even scan lines. There are 60 fields in one second of video.

filter—An electrical circuit designed to prevent the passage of certain frequencies.

flagging—Bending at the top of a picture played back by a VCR.

footprint—The signal coverage of a satellite.

frame—One complete video picture, comprising both odd and even fields. There are 30 video frames per second.

frequency response—As used in audio applications, sonic range; the highest and lowest audio frequencies that can be accurately reproduced. As used in video applications, brightness range; the darkest and lightest tones that can be accurately reproduced.

gain—Amplification, usually expressed in dB.

GLOSSARY

ghost—A duplication of the video image (a kind of visual echo), usually caused by a reflection in the broadcast signal or an improperly terminated signal input or output.

gigahertz (GHz)—A measurement of frequency. One gigahertz is equal to one billion cycles per second.

ground—Refers to the point of (usually) zero voltage, and can pertain to a power circuit or a signal circuit.

guard band—A blank area on the tape separating two signal tracks.

HQ—A set of circuits used in some VHS video cassette recorders that provide improved video resolution. The HQ "standard" is a set of four separate picture-improvement circuits (true HQ), though some VCRs only use one or two of the circuits (mock HQ).

helical scan—The technical name for the way the video heads in a VCR record and play back picture information. Also used to record and play back stereo hi-fi audio.

hertz—Abbreviated Hz. A unit of measurement used for expressing cycles per second, named after German physicist H.R. Hertz. One hertz equals one cycle per second. Also used with the letters "k," "M," and "G" as multipliers to indicate thousands, millions, and billions, respectively.

heterodyne—See *color-under*.

hi-fi—A general term used to denote the capability of reasonably high-fidelity audio playback. Hi-fi VCRs are equipped with separate rotating audio heads (VHS) or mix sound and picture in one signal and record both simultaneously on tape (Beta).

hiss—Audible high-frequency noise. Can be caused by the recording medium (such as video tape) or circuitry.

hum—A low-level, low-frequency electrical noise usually caused by the alternating current in household power lines interfering with audio and video signal circuits. Hum is often picked up in cables extending between the VCR and TV or hi-fi but can also be caused by an ungrounded circuit. The hum is audibly apparent as a low-frequency (60 Hz) tone; video "hum" is a dark bar that floats through the picture.

IC—Integrated circuit. Also called a chip. A complete electrical circuit housed in a self-contained package. See also *LSI*.

IR—Short for *infrared*, used in VCRs for remote sensing and remote control applications.

idler—A wheel most often used to drive the supply and take-up tape reel spindles. The idler is powered by a separate motor (rarely) or driven by a belt connected to the capstan motor.

impedance—The degree of resistance that an alternating electrical current (ac) will encounter when passing through a circuit, device, or wire. The amount of impedance is expressed in ohms.

impedance roller—Metal or plastic rollers used in most VCRs to provide an even and steady flow of tape through the transport mechanism.

insertion loss—The attenuation or loss of power caused by the insertion of a component or circuit into the signal path.

interlace—The process of combining the odd and even lines of subsequent video fields to generate a complete frame.

jack—A term used for the cable connector in audio and video equipment.

Ku band—The 12 GHz frequency band used for some satellite television transmission.

Kelvin (degrees)—A temperature measurement used in grading low-noise amplifiers, as used in satellite television systems.

kilohertz (kHz)—A measurement of frequency. One kilohertz is equal to one thousand cycles per second.

LED—Short for *light emitting diode*, a unique type of semiconductor that is made to emit a bright beam of light. Often used as a panel indicator, but is also employed in remote sensing systems and infrared remote control devices.

LNB (Low Noise Block down-converter)—In satellite TV systems, the part that receives the signal and amplifies it for use by the receiver. An LNB contains both a low-noise amplifier and block down-converter.

LP—Long play, the intermediate recording speed found on many (but not all) VHS VCRs. See also *SP* and *EP*.

LSI—Large scale integration. A complex integrated circuit (IC) that is a combination of many IC's and other electronic components that are normally packaged separately. LSI (and VLSI, for Very Large Scale Integration) chips are used in the latest video equipment and held proprietary by the manufacturer.

loading—A function of the VCR where the cassette is drawn into the machine and readied for threading. Loading is automatic in front-load VCRs and manual for top-load VCRs.

logic—Primarily used to indicate digital circuits or components that accept one or more signals and act on those signals in a pre-defined, orderly fashion.

luminance—A term used to denote the brightness or black and white picture of a video image.

LV—Short for *LaserVision*, the trade name of optical video discs.

MOS—Short for *metal-oxide semiconductor*, a type of imaging device used in certain video cameras and camcorders.

MTS—An acronym for *multi-channel television sound*, the stereo broadcasting standard used in the U.S.

M-load—The tape threading "pattern" used by 8mm and VHS VCRs. Also see U-load. The term is derived by the "M" shape of the tape when threaded in the VCR transport mechanism.

matching transformer—A device used to match the impedance between one cable and device to another cable or device. Matching transformers are most often used to balance the impedance of an outdoor antenna (usually 300 ohms) to the impedance of modern TV sets and VCRs (usually 75 ohms).

megahertz (MHz)—A measurement of frequency. One megahertz is equal to one million cycles per second.

micron—A unit of measurement; one micron (or 1μ) is equal to 0.001mm.

microprocessor—A special integrated circuit that performs semi-intelligent functions based on instructions written in a program. Fundamentally, a microprocessor is a hardware device that can be electrically "rewired" by using new software. Often considered the "brain" of a computer or circuit. Many of the later VCRs use microprocessors to control system functions.

mid-band—The frequencies found between channels 6 and 7, designated channel A through I. Often used in cable television systems.

modulation—A way in which one signal modifies or controls another signal for such purposes as enabling it to carry information. Often used to describe radio frequency (rf) transmission. RFM is frequency modulation; AM is amplitude modulation.

GLOSSARY

monitor—A video display. A monitor is like a TV, but it lacks the ability to tune in channels. A monitor may or may not have a sound amplifier and speaker.

monitor/TV—A combination monitor and television set.

noise—An unwanted signal. Audio noise is usually heard as hiss; video noise is usually seen as "salt and pepper" flecks.

NTSC—National Television Standards Committee. A group of businesses and engineers originally created to decide on early standards for color and black and white television in the U.S. The NTSC system is also used in Japan. Other television standards around the world include PAL (most of Europe) and SECAM (France, parts of Africa, and the Soviet Union).

ohm—The unit of measure of impedance or resistance.

oxide—The magnetic coating applied to video tape that actually stores the video and audio signals generated by the VCR.

PCM—Short for *pulse code modulation*, a way of digitally recording an audio signal. Used in most 8mm tabletop decks and in some high-end VHS recorders.

phone plug—A certain type of cable connector often used for headphone and microphone audio applications. It has a single stem, with parts of the stem insulated to provide for additional contact(s). Phone plugs come in many sizes: ¼-inch phone plugs are large and used for headphones and some microphones; ⅛-inch (or miniature) phone plugs are smaller and used mainly for lightweight headsets. See also *phono plug*.

phono plug—A certain type of cable connector used in home audio and video applications. It is composed of a round center stem and an insulated metal cap. Also called an RCA plug. See also *phone plug*.

pickup tube—The photo-receptor of a video camera.

pinch roller—A rubber roller that presses against the capstan to pull the tape through the transport mechanism.

RAM—Random access memory. As used in VCRs, electronic memory used to temporarily store programming or picture information.

rf—Radio frequency. In television, vhf, uhf, and cable TV signals are considered rf.

rf modulator—A device that adds a carrier to baseband audio and video so the signal can be picked up by a VCR or TV.

resistance—Opposition to direct electrical current (dc), expressed in ohms.

resolution—The clarity of sharpness of the picture. Resolution is most often stated in the number of total lines that make up an image or in megahertz (millions of cycles per second).

roller—A rubber tire or wheel.

SAP—Short for *special audio program*, a separate sound signal used in some multi-channel television sound (MTS) broadcasts.

SP—Short play (or standard play), the fastest speed setting in VHS VCRs. See also *LP* and *EP*.

scanner—The video head drum assembly.

separation—The complete electrical separation of two or more signals that usually refers to either right and left stereo channels, audio and video signals, or luminance and chrominance signals.

servo—An electronic circuit that modifies its output in accordance to a constantly varying input signal.

shell—A cassette tape cartridge.

shielding—A conductive material surrounding but insulated from an electrical conductor.

signal—The desired portion of electrical information.

signal-to-noise ratio (S/N)—The ratio of wanted signal to unwanted signal, usually expressed in dB. The higher the number, the better.

splitter—A device that separates a signal into two (or more) signal paths of substantially equal strengths.

stereo adaptable—A VCR or TV that can be made to receive MTS stereo audio broadcasts by the addition of an external decoder.

stereo ready—A VCR or TV that already has the necessary circuitry to decode MTS stereo audio broadcasts.

subcarrier—A frequency-derived signal of a main signal.

Super VHS—A superset of the VHS format that enables recording and playback of very high resolution video.

supply reel spindle—The left spindle in the VCR that drives the supply tape reel of a cassette.

sync—Short for *synchronization*, a broad term to indicate the proper ordering of electrical signals to generate and display sound and picture from a video tape.

take-up reel spindle—The right spindle in the VCR that drives the take-up tape reel of a cassette.

tape guide spindle—One of several posts used in VHS and Beta VCRs to accurately position the tape in the transport mechanism.

tape-load switch—A small leaf switch used in some VCRs to detect when the threading mechanism has fully threaded the tape around the video heads.

tape tension lever—A lever that is used to detect tape supply reel tension. The lever is connected to a brake; when the tension decreases, the brake tightens to slow the reel. The reverse occurs when the tape tension increases.

terminating resistor—A resistor (usually 75 ohms) attached to the end of a cable or to an input or output on a piece of video equipment. The resistor restores proper system impedance.

threading—A function of the VCR where tape is withdrawn from the cassette and placed against the rollers and magnetic heads, in order to facilitate recording and playback. Threading is accomplished by the threading mechanism.

threading ring—The threading mechanism used in Beta VCRs.

torque—The amount of mechanical force exerted on an object. In VCRs, torque usually relates to the rotational force exerted by the supply and take-up reels.

track—1. One of several discrete signals applied to video tape during recording. 2. Parallel bands of video information recorded in a diagonal on the video tape.

tracking—The alignment of the video heads in a VCR to tracks already recorded on the tape.

tracking control—An electronic control that varies the tracking timing of the VCR.

transponder—A channel on a satellite.

trap—A filter.

twinlead—TV hookup cable that consists of two wires, separated a certain distance by a plastic spine.

U-load—The tape threading "pattern" used by Beta VCRs. Also see *M-load*. The term is de-

GLOSSARY

rived by the "U" shape of the tape when threaded in the VCR transport mechanism.

uhf—Ultra high frequency, TV channels 14 through 83, occupying the frequency spectrum of 300 MHz to 3,000 MHz.

VCR—An acronym for video cassette recorder.

VCP—An acronym for video cassette player.

vhf—Very high frequency, TV channels 2 through 13, occupying the frequency spectrum of 30 MHz to 300 MHz.

VHS—A VCR format pioneered by JVC.

video head drum—The cylindrical-shaped mechanism that houses the video heads (it may also house the hi-fi audio heads in a VHS hi-fi VCR).

writing speed—The effective video head-to-tape speed during playback and recording. The writing speed considers the rotational speed of the video heads, not just the linear speed of the tape as it travels through the transport mechanism of the VCR.

Index

Index

8mm format VCRs, 2-3
  pulse code modulation (PCM)
    audio in, 59
  recording and playback on, 43

A
A-B switch, 71
ac leakage, 253
accessories, 82
adjustments
  control/audio head, 182
  erase head tape post, 184
  tape guide spindle, 180
attenuator, 72
audio signals, recording and
  playback of, 42
audio dubbing, 16
audio heads, 5, 6
audio I/O terminals, 241
audio/video terminals, 236
automatic color control (ACC), 42
automatic phase control (APC),
  42

B
Baird, John L., 19
balun, 69

band splitter, 71
baseband signals, 25
batteries, 264
bearings, 251
belt tire, 213
belts, 121
  maintenance of, 117
Beta format VCRs, 2
  hi-fi audio in, 51
  cassette for, 3
  misthreading in, 259
  recording and playback on,
    43-45
  tape threading in, 27
BII speed, 14
BIII speed, 14
brakes, 217
  maintenance of, 123
brightness noise reduction, 17
brightness/contrast controls (TV),
  161
burst signal, 22

C
cable and component signal loss,
  309-310
cable-ready VCRs, 12

cables, 66, 69, 189, 238-244
  care of, 110
  shorted, 254
  troubleshooting of, 165
camcorder, 5, 6
capstan, 33, 34, 121, 214, 219,
  227
  drive belts for, 244, 250
capstan pinch roller, 219, 227,
  244
  maintenance of, 117
capstan roller rejuvenator lathe,
  104
cassette loading switch, 26, 201
cassette will not eject, 205-208
cassette will not load, 200-204
cassette-in switch, 198, 229
cassette-lift mechanism, 202
cathode ray tube, 20
chroma signal, 22
  conversion of, 38, 40
cleaner/degreaser, 101, 125
cleaning and preventative
  maintenance, 107-118
cleaning supplies, 103
coaxial cables, 66
color flashes on and off, 258

color noise reduction, 17
color processors, 84
color signal, 22
color television, 24
concealed damage, 64
connectors, 61, 68
    F-, 306-308
    printed circuit board, 197
continuity testing, 111, 176
control motor/solenoid, 222
control track head, 9, 10, 33, 238, 245
control/audio heads, adjustments to, 182
controller IC, 224
controls, 60
    cleaning of, 136
    connecting wires to, 196
CTL systems, 18
cue, 15

D
dc motor testing, 204
debugging, 75, 77
decoders, 85
demagnetizing, 134
detail enhancer, 17
detent tuners, 11
dew sensor, 196, 197, 213, 216, 229
digital effects VCRs, 18, 58
disassembly, 112
Dolby, 10
dots, dashes, wavy lines, 164
drop-out, 263
drop-out compensator (DOC), 42
dropped VCRs, 149
    broken wires in, 150
    internal damage to, 150
    leakage current test for, 151
dubbing, 10
    audio and video, 16

E
eject switch, 207
electric shocks from VCR, 252-255
electrical contact cleaning, 126
electron beam scanning, 21, 24
electronic tuners, 11
electrostatic discharge, 304
emergency repairs, 149-159
    dropped VCR, 149
    fire damaged VCRs, 153
    foreign object removal, 157
    leaked batteries, 159

sand, dirt, dust damaged VCRs, 157
smoke damaged VCRs, 154
tools and safety for, 149
water damaged VCRs, 155
Emerson VCRs, 266
end-of-tape sensor, 135, 198
    bent or broken, 260
enhanced VCRs, 15
erase heads, 5
    adjustment to, 184
extended play speed, 14
exterior dust and dirt, 109

F
F-connectors, 306-308
Farnsworth, Philo T., 20
fast-forward button, 13
fast-forward or rewind won't operate, 220-222
FG tachometer pulses, 34
fields, 23, 24
filters, 72, 245
final inspection, 137
fine tuning (TV), 160
fire damaged VCRs, 153
Fisher VCRs, 266-267
flagging, 261
flowcharts, 170, 186
flutter, 33
flux field, 37
flying erase heads, 50, 50
flywheel, 34, 121
foreign objects in VCRs, 157
formats, compatible, 225
frames, 23, 24
freeze frame, 15
frequency meter, 95
frequency modulation (FM), 22
frequency spectrum, television, 311-315
front panel control(s) inoperative, 228-230
front panel indicators inoperative, 231-232
full-erase head, 6
function controls, 13
function switch, 222
fuses, blown, 189, 257

G
GE VCRs, 268
gears, 250
ghosts, 164
grounding rod, 73
guard bands, 49

H
Hall-effect ICs, 31
hashed picture, 49
head cylinder, 8
head drum, 8, 31, 116
    propulsion units for, 31
head gap size, 43
heads
    audio, 5, 6
    cassette cleaners for, 127
    control track, 9, 10, 33, 238, 245
    dirty, 47, 238, 241, 244
    erase, 5
    flying erase, 11, 50
    full-erase, 6
    magnetic, 37
    maintenance of, 126
    manual cleaning of, 129
    playback, 37
    recording, 37
    rotating audio, 10
    simplified view of, 7
    special effects, 224
    stereo audio, 9
    switching, 51, 52
    three- and five-, 49
    video, 5, 8, 48
    worn, 239
headwheel, 8
heterodyning, 40
Hi-Fi audio, 51
    limitations of, 55
Hitachi VCRs, 268-271
hookup, 66
    parts for, 69
horizontal sync pulse, 23
HQ circuits, 17

I
I/O terminals, 254
idler tire, 122, 216, 221
idler wheel, 217
image enhancer, 83
index address, 18
information tracks, 7
infrared detector, 96
    circuit description of, 98
    use of, 100
input/output terminals, 15
insertion loss, 83
installation, 65, 257
    achieving optimum picture and sound during, 78
    checkout after, 75
    debugging, 77

334

matching impedances in, 67
tips for, 74
internal components, topical
cleaning of, 124

J

jacks, 241
JVC VCRs, 271

L

LCDs and LEDs, 231
leaky battery damage, 159
linear sound tracks, 42
linear tape speed, 9
lines, 23
Lloyd's VCRs, 271
loading motor, 202, 206, 208
logic probe, 91
shaft encoder testing with, 262
logic pulser, 92
making your own, 93
long play speed, 14
lubrication, 103, 131
luminance signal, 22, 23
conversion of, 38, 40

M

Magnavox VCRs, 272
magnetic flux field, 37
maintenance log, 137, 297-301
manuals, 109
service and repair, 179
matching impedances, 67
matching transformers, 69
Mitsubishi VCRs, 272
monitor, 100
Multi-Tech VCRs, 273
multiplex (MPX) jack, 16

N

NEC VCRs, 273
noise bars, 48
non-VCR troubleshooting,
160-168
numeric address, 18

O

oiling, 131
operating controls, 13
operational manuals, 109
opposing azimuth, 48
oscilloscope, 94

P

packing lists, 65

Panasonic VCRs, 274
pause button, 13
PCM audio, 59
PG timing pulses, 34
picture, optimum, 78
picture tubes, 23
play button, 13
playback
audio signal, 42
heads for, 37
signal conversion for, 40
sparkles in, 259
speeds of, 3
weak video and audio, 259
PortaVideo VCRs, 276
power cords, polarity of, 76
power supply, 119, 189, 201,
206, 230, 255, 257
schematic of, 193
preliminary inspection, 115
prescaler, 96
pressure roller, 218
printed circuit board, 118, 198,
224, 231, 242, 255, 257
connector to, 197
program reception
troubleshooting, 161
programmable VCRs, 12
pulse width modulation, 32, 34

Q

Quasar VCRs, 277

R

Radio Shack/Realistic VCRs, 279
RCA VCRs, 279
reassembly, 137
record button, 13
recording
audio signal, 42
inoperative, 257
signal conversion for, 38
recording heads, 37
reel motor, 222
reel spindles, 135
reel tire, 213
reel-drive motor, 218
reel-drive tire, 224
remote control, 13, 229
cleaning of, 138
inoperative, 246-248
interference in, 166
placement of, 66
repair manuals, 179
review, 15
rewind button, 13

rewind/fast forward mechanisms,
221
rf modulator, 83, 120, 236
RF signals, 25
RF switcher, 82
rotating audio heads, 10

S

safety precautions, 108
Samsung VCRs, 281
sand, dirt, and dust damaged
VCRs, 157
Sanyo VCRs, 282
scanner, 8
scanning lines, 15, 21, 24
search inoperative, 223-224
search switches, 224
shaft encoder, 262
Sharp VCRs, 284
shielding, 242
signal amplifier, 73
signal conversion, 38
signal loss, 309-310
signal splitter, 70
slow motion modes, 15
smoke damaged VCRs, 154
snowy video, poor audio,
243-245
soldering tips and techniques,
302-305
Sony VCRs, 284
sound
optimum, 79
good/video poor, 237-239
sources, 291-293
special effects
audio, 15
heads for, 224
video, 15
specifications charts, 316-322
spilled tapes, 207
splicing tapes, 142
spray cleaner, 101
standard play speed, 14
static electricity build-up, 114
stereo adaptable/ready VCRs, 15
stereo audio heads, 9
stereo synthesizers/decoders, 85
stretched tapes, 141
suggested reading, 294-296
Super Beta format, 17
recording and playback on, 56
Super VHS format, 18
recording and playback on, 56
supply reel spindle, 36, 216
supply reels, 225

switch selector (TV), 160
switches, 221
  front panel, 229

T

take-up reel sensor, 214
take-up reels, 225
take-up spindle, 35, 216
tape counter, 261
tape guide spindles, 210, 217, 225
  adjustment to, 180
tape indicator, 201
tape loading mechanisms, 13, 250
tape speed, capstan role in, 33
tape threading mechanisms, 27, 250
tape transport, 31
television
  baseband vs. RF signals, 25
  development of color, 24
  first commercial networks for, 20
  frequency spectrum for, 22, 311-315
  history of, 19
  lines, fields, and frames in, 23
  operational theory of, 21
  picture tube of, 23
  signal breakdown in, 22
television troubleshooting, 160-165
  flickering picture in, 161
  isolating reception problems in, 161
temperature differences, 196
test tapes, 105
testing, 75
threading mechanisms, 210
threading motor, 211
threading ring, 211
time shifting, 12
timer inoperative, 233-234
timer mode, 189, 206
tires
  idler, 122
  maintenance of, 117
tools and supplies, 86-106
torque limiters, maintenance of, 123
Toshiba VCRs, 286
tracking control
  inoperative, 225-227
  troubleshooting of, 165
transformers, 119, 254

primary and secondary, 190
troubleshooting
  determining repairable problems during, 170
  flowcharts for, 170
  miscellaneous difficulties, 257
  techniques for, 174
troubleshooting charts, 187-264
  cassette will not eject, 205-208
  cassette will not load, 200-204
  electric shocks, 252-255
  fast-forward or rewind won't operate, 220-222
  front panel control(s) inoperative, 228
  front panel indicators inoperative, 231-232
  remote control inoperative, 246-248
  search inoperative, 223-224
  snowy video, poor audio, 243-245
  sound good/video poor, 237-239
  timer inoperative, 233-234
  tracking control inoperative, 225-227
  TV tuner channels don't change, 235-236
  VCR does not turn on, 188-194
  VCR eats tape, 215-219
  VCR noises, 249-251
  VCR overheats, 256-257
  VCR turns on/no operation, 195-199
  VCR will not play tape, 212-214
  VCR will not thread tape, 209-211
  video good/sound poor, 240-242
troubleshooting, 169-177
tuner, 11, 236
  cleaning, 134
TV tuner channels don't change, 235-236
TV/VCR switch, 236
  troubleshooting, 165
twinlead cable, 68

U

U-load technique, 28
unpacking, 64
unthreading, 31

V

VCR does not turn on, 188-194

VCR eats tape, 215-219
VCR malfunctions, 178-264
VCR noises, 249-251
VCR overheats, 256-257
VCR turns on/no operation, 195-199
VCR will not play tape, 212-214
VCR will not record, 257
VCR will not thread tape, 209-211
ventilation, 65, 257
vertical hold control (TV), 161
vertical sync pulses, 23, 43
VHS format, 2
  cassette for, 3
  head geometries of, 47
  hi-fi audio in, 54
  playback circuit block diagram for, 41
  recording circuit block diagram for, 39
  tape threading in, 30
  track spacing on, 46
video cassette players (VCPs), 5
video cassette recorders
  accessories, 82
  basic block diagram of, 25, 26
  basic operation of, 13
  cable ready capabilities of, 12
  cassette loading in, 26
  cleaning and preventative maintenance of, 107-118
  connectors for, 61
  control track head in, 9
  controls for, 60
  digital effects (computerized), 18
  enhanced, 15
  first aid for, 149-159
  formats for, 2
  information tracks on, 7
  input/output terminals on, 15
  installation of, 64-85
  installation troubleshooting guide for, 78
  introduction to, 1-18
  maintenance tools and supplies for, 86
  malfunction troubleshooting, 178-264
  operating controls on, 13
  operation of, 19-64
  playback speeds of, 3
  preventative maintenance schedule for, 147
  programmable, 12
  proper tape handling for, 80

Other Bestsellers From TAB

☐ **MAINTAINING AND REPAIRING VCRs—2nd Edition—Robert L. Goodman**

"... of immense use ... all the necessary background for learning the art of troubleshooting popular brands" said *Electronics for You* about the first edition of this indispensable VCR handbook. Revised and enlarged, this illustrated guide provides complete, professional guidance on troubleshooting and repairing VCRs from all the major manufacturers, including VHS and Betamax systems and color video camcorders. Includes tips on use of test equipment and servicing techniques plus case history problems and solutions. 352 pp., 427 illus.

Paper $17.95 **Hard $27.95**
Book No. 3103

☐ **TROUBLESHOOTING AND REPAIRING AUDIO EQUIPMENT—Homer L. Davidson**

When your telephone answering machine quits ... when your cassette player grinds to a stop ... when your TV remote loses control ... or when your compact disc player goes berserk ... You don't need a degree in electronics or even any experience. Everything you need to troubleshoot and repair most common problems in almost any consumer audio equipment is here in a servicing guide that's guaranteed to save you time and money! 336 pp., 354 illus.

Paper $17.95 **Hard $25.95**
Book No. 2867

☐ **SATELLITE COMMUNICATIONS—2nd Edition— Stan Prentiss**

This revised, updated, and expanded edition of a bestseller provides a comprehensive, easy-to-follow look at satellite technology as it exists today, with special emphasis on television receive-only earth stations (TVROs) and signals to them from geosynchronous orbiting satellites. Includes up-to-date nomograph charts to aid you in locating any satellite from your own specific longitude and latitude information on TVRO dishes, feeds, receivers, and special space noise/loss/EIRP, and uplink transmissions. 320 pp., 152 illus.

Paper $16.95 **Hard $21.95**
Book No. 2792

☐ **TROUBLESHOOTING TECHNIQUES FOR MICROPROCESSOR-CONTROLLED VIDEO EQUIPMENT—Bob Goodman**

With this excellent introduction to servicing these "electronic brains" used in everything from color TVs and remote control systems to video cassette recorders and disc players, almost anyone can learn how to get to the heart of most any problem and solve it skillfully and confidently. Includes dozens of handy hints and tips on the types of problems that most often occur in microprocessors and the easiest way to deal with them. 352 pp., 232 illus.

Paper $16.95 **Hard $24.95**
Book No. 2758

☐ **THE ILLUSTRATED HOME ELECTRONICS FIX-IT BOOK—2nd Edition—Homer L. Davidson**

This revised edition of the bestselling home electronics fix-it handbook will save you time and aggravation AND money! It is the only repair manual you will ever need to fix most household electronic equipment. Packed with how-to illustrations that any novice can follow, you'll soon be able to fix that broken television and portable stereo/cassette player and "Boom Box" and intercom and ... the list goes on! 480 pp., 377 illus.

Paper $16.95 **Hard $25.95**
Book No. 2883

☐ **THE CAMCORDER HANDBOOK—Gerald V. Quinn**

Takes you step-by-step through the entire production process using a camcorder. Going well beyond what is included in the standard owner's manual, Quinn provides guidance in lighting, sound, camera movement, and more—for shooting indoors or out! Whether you're taping a school play, sporting event, wedding, vacation, children, recording a family history, or making a tape for business purposes, you'll have the basic understanding of video concepts and techniques that you need. 240 pp., 139 illus.

Paper $12.95 **Hard $18.95**
Book No. 2801

☐ **COMPACT DISC PLAYER MAINTENANCE AND REPAIR—Gordon McComb and John Cook**

Packed with quick and reliable answers to the problems of maintaining and repairing CD players, this illustrated, do-it-yourself guide takes the apprehension out of first-time repairs. The authors take away the mystery that surrounds these seemingly complicated devices and give you the confidence you need to repair minor malfunctions (the cause of more than 50% of CD player problems). 256 pp., 188 illus.

Paper $13.95 **Hard $19.95**
Book No. 2790

☐ **TROUBLESHOOTING AND REPAIRING SOLID-STATE TVs—Homer L. Davidson**

Packed with case study examples, photos of solid state circuits, and circuit diagrams. You'll learn how to troubleshoot and repair all the most recent solid-state TV circuitry used by the major manufacturers of all brands and models of TVs. This workbench reference is filled with tips and practical information that will get you right to the problem! 448 pp., 516 illus.

Paper $17.95 **Hard $26.95**
Book No. 2707

Other Bestsellers From TAB

☐ **GETTING THE MOST OUT OF YOUR VIDEO GEAR—Gerald V. Quinn**

Much more than just a catalog or owner's manual, this guide addresses the real needs of the home video user. It covers using and purchasing equipment, supplies, and accessories plus advanced topics like image enhancers, color processors, and audio mixers. Microphones and lighting, production, planning techniques, strategies used by the professionals, career ideas—it's all here! 256 pp., 188 illus.

Paper $12.95   **No. 2641**

☐ **VIDEO PRODUCTION—THE PROFESSIONAL WAY—Carl Caiati**

Now, right in step with the latest techniques and equipment available to the video recording world, this hands-on sourcebook provides an in-depth introduction to professional-quality video tape production that no video enthusiast can afford to miss! Caiati provides guidance in all the basic taping and editing procedures, plus tips on equipment maintenance and troubleshooting. 256 pp., 306 illus.

Paper $16.95   **Hard $24.95**
Book No. 1915

☐ **TROUBLESHOOTING AND REPAIRING SATELLITE TV SYSTEMS—Richard Maddox**

This first-of-its-kind troubleshooting and repair manual can mean big bucks for the professional service technician (or electronics hobbyist looking for a way to start his own profitable servicing business) . . . *plus big savings for the TVRO owner who wants to learn more about maintaining and servicing his own receiver!* Includes service data and schematics! 256 pp., 479 illus.

Hard $26.95   **Book No. 1977**

☐ **BEGINNER'S GUIDE TO TV REPAIR—3rd Edition—Homer L. Davidson**

Now, anyone can locate and correct dozens of common TV problems quickly, easily, and inexpensively! You'll learn how to keep both black-and-white and color TV sets performing their best . . . to remedy problems that cause loss of vertical, loss of picture, or loss of both sound and picture. Color set problems from burst amplifiers to color sync and color control circuits are also covered. 272 pp., 82 illus.

Paper $14.95   **Book No. 1897**

Send $1 for the new TAB Catalog describing over 1300 titles currently in print and receive a coupon worth $1 off on your next purchase from TAB.

(In PA, NY, and ME add applicable sales tax. Orders subject to credit approval. Orders outside U.S. must be prepaid with international money orders in U.S. dollars.)
Prices subject to change without notice.

To purchase these or any other books from TAB, visit your local bookstore, return this coupon, or call toll-free 1-800-233-1128 (In PA and AK call 1-717-794-2191).

Product No.	Hard or Paper	Title	Quantity	Price

☐ Check or money order enclosed made payable to TAB BOOKS Inc.

Charge my   ☐ VISA   ☐ MasterCard   ☐ American Express

Acct. No. _____ Exp. _____

Signature _____

Please Print
Name _____

Company _____

Address _____

City _____

State _____ Zip _____

Subtotal	
Postage/Handling ($5.00 outside U.S.A. and Canada)	$2.50
In PA, NY, and ME add applicable sales tax TOTAL	

Mail coupon to:
TAB BOOKS Inc.
Blue Ridge Summit
PA 17294-0840

BC

recording and playback operations in, 37
reference guide for, 265-287
remote control, 13
specifications charts for, 316-322
tape threading in, 27
tape transport in, 31
troubleshooting for, 169-177
tuning of, 11
types of, 14
video and audio heads in, 5
video dubbing, 16
video furniture, 65
video good/sound poor, 240-242
video head drum, 116
obstructions to, 214
video heads, 5
geometries of, 43
maintenance of, 126

number of, 9
rotational head drum of, 8
video preamplifier, 239
video processing circuits, 239
video stabilizers, 84
videocassettes
compatible formats for, 225
cover release pin on, 203
cover release pins on, 143
damaged, 81, 199, 202, 213, 216
handling of, 80
jammed tapes in, 142
life of, 80
maintenance of, 141
spilled tapes in, 207
splicing tapes in, 142
tape shell damage in, 144
troubleshooting, 166
volt-ohm meter, 87

accuracy and functions of, 89
automatic ranging in, 89
digital vs. analog, 88
resistance test with, 191
safety and use of, 90
testing indicator lights with, 231
use of, 175
voltage regulator, 194

W
water damaged VCRs, 155, 254
white clip level enhancement, 17
wiring and circuitry, 238, 254
workspace, 86
wow, 33
writing speed, 9

Z
Zenith VCRs, 287
Zworykin, Vladimir, 20